Additional praise for EINSTEIN'S TELESCOPE

"Dark matter and dark energy rule the Universe, if our new picture of the world is correct. Evalyn Gates is your intrepid guide into these dark mysteries, shining the light of today's astronomical evidence on WIMPS and MACHOS, cosmological acceleration, and warped space. . . . There's a scientific revolution in the making in the dark cosmos, and Evalyn Gates gives the interested reader a clear idea of what's at stake."
—Robert P. Kirshner, Clowes Professor of Science, Harvard University, and author of *The Extravagant Universe*

"In *Einstein's Telescope*, Evalyn Gates, an expert on all that's dark in the universe, brings dark matter, dark energy, and even black holes to light. With deft, humor, and grace, she tells us why we know they're there, how we are trying to see them, and what it could mean for our understanding of the cosmos if we did."
—Neil deGrasse Tyson, astrophysicist, American Museum of Natural History, and author of *Death By Black Hole* and *The Pluto Files*

"An enthusiastic update on the search for the materials that make up the universe. . . . Splendidly satisfying reading."
—*Kirkus Reviews*, starred review

"A testament to what can be done when beautiful mathematics is applied to astronomy. . . . Gates is . . . well-equipped to come up with clear, well-thought-out explanations of the various areas in which lensing is used. . . . All in all, this is a story of teasing out improbable, almost indiscernible signals to great effect. . . . Gates uses one of Einstein's great ideas to weave together mathematics, astrophysics, and cosmology into a coherent and, dare I say it, compelling narrative that maps out the frontiers of contemporary research."
—Pedro Ferreira, *Science*

"Gates offers clear, accessible explanations of how gravitational lensing can be used to solve the biggest mysteries of the universe—first by using luminous matter to ferret out the dark matter, then by using the cosmic web of dark matter itself as a lens to probe dark energy and the very structure of time and space." —*New Scientist*

"Gates does a masterful job of making usable examples and explanations. . . . *Einstein's Telescope* provides a direction for the future, new telescopes and experiments to come, that may even better answer more questions than it raises." —*Sacramento Book Review*

"Evalyn Gates is an active scientist with a background in particle physics, cosmology, and public outreach. This combination gives her the ideal tools to describe gravitational lensing. She addresses the book to scientifically curious laypeople—and delivers. Guiding the reader through multiverses, accelerated cosmic expansion or the physics of exploding stars, Gates clearly explains the problems and challenges that physicists face today. She uses apt analogies and original comparisons, many of which I had not heard before. . . I found myself eagerly awaiting the next good metaphor. . . . Through Gates's ambitious book, everyone will appreciate the puzzles of the dark Universe—and the power and beauty of gravitational lensing."
—Joachim Wambsganss, *Nature*

"[Gates] succeeds in presenting mind-boggling ideas with such an engaging, readable style. . . . Appreciate the sheer beauty of the book's astronomical images as . . . [you] learn how the images were obtained and what they represent. Gates writes with a freshness and clarity that make complex ideas such as relativity, lensing, black holes, and the cosmic web understandable. Highly recommended." —*Library Journal*

"Cosmology has been upended by the discoveries of dark matter and dark energy. This guide to the tumult comes from a talented scientist in astrophysics who has a gift for presenting the topic to a nonscientific audience. . . . Supplemented by beautiful astronomical photo-

graphs and instructive illustrations . . . [*Einstein's Telescope* is] as exciting as it is informative."
—Gilbert Taylor, *Booklist*, starred review

"A highly accessible account of the modern approach to an ancient question: 'What is the nature of the Universe?' . . . Gates gives a cutting-edge account of how gravity rules the Universe, and what astronomers have discovered by seeing gravity in action on a gigantic cosmic scale. The book's title refers to gravitational lensing. . . . According to Gates, a cluster of galaxies can produce beautifully bizarre, almost kaleidoscopic, images from a lens sculpted out of spacetime. *Einstein's Telescope* brilliantly summarizes half a century of progress in which physical cosmology moved from being just a handful of crudely determined numbers to a precision science."
—Simon Mitton, *Times Higher Education*, "Book of the Week"

"Dr. Gates succeeds in navigating the reader through the complexities of general relativity and quantum mechanics to arrive at our contemporary understanding of the universe and its challenging questions. She does this in a straightforward and clear manner that makes *Einstein's Telescope* the perfect book for a non-specialist to gain essential insights into modern cosmology. In doing so, she prepares the reader for the next great revolution in science as we follow the search for dark matter and dark energy."
—Paul H. Knappenberger, Jr., PhD, president,
Adler Planetarium & Astronomy Museum

"What stands out in Evalyn Gates' cogent review of this intriguing topic is the sheer cleverness astronomers have demonstrated in fashioning tools to study the unseeable. . . . As Gates so aptly demonstrates, describing how a science works toward a solution provides . . . enjoyment for any reader."
—Marcia Bartusiak, *Washington Post Book World*

"People are always asking me how scientists can be so sure that most of the universe is made of mysterious-sounding stuff, dark matter and dark energy. Now I can just hand them Evalyn Gates's book. *Einstein's Telescope* takes a fantastic story from modern cosmology and makes it down-to-earth, accessible, and fun."

—Sean Carroll, senior research associate in physics, California Institute of Technology, and author of *Spacetime and Geometry: An Introduction to General Relativity*

"A solid, easy-to-understand introduction to both gravity lensing and the search for dark matter. . . . [Gates] writes in clear, concise prose . . . like the late astronomer Carl Sagan, she treats her readers as intelligent, educated people who happen to not be experts in a very specialized field of study."

—Jim Trageser, *North County Times*

EINSTEIN'S TELESCOPE

The Hunt for Dark Matter and Dark Energy in the Universe

EVALYN GATES

W. W. Norton & Company

New York London

To Eric,
Greg, Jordan, and Kendall

Printed in the United States of America
First published as a Norton paperback 2010

Manufacturing by RR Donnelley, Bloomsburg
Book design by Soonyoung Kwon
Production manager: Anna Oler

Library of Congress Cataloging-in-Publication Data

Gates, Evalyn.
Einstein's telescope : the hunt for dark matter and
dark energy in the universe / Evalyn Gates.
p. cm.
Includes bibliographical references and index.
ISBN 978-0-393-06238-0 (hardcover)
1. Dark matter (Astronomy) 2. Dark energy (Astronomy) I. Title.
QB791.3.G38 2009
523.1'126—dc22 20080444455

ISBN 978-0-393-33801-0 pbk.

W. W. Norton & Company, Inc.
500 Fifth Avenue, New York, N.Y. 10110
www.wwnorton.com

W. W. Norton & Company Ltd.
Castle House, 75/76 Wells Street, London W1T 3QT

1 2 3 4 5 6 7 8 9 0

Contents

Preface

Cosmology—the study of the Universe and its evolution—is in the midst of a truly phenomenal era. This book was written as an invitation, extended to anyone and everyone interested in the incredible ideas and images that are shaping and reshaping our understanding of the cosmos. It is offered in the same spirit that musicians proffer their music to the world—to be absorbed in many different ways at many different levels. Science is too often segregated into one small box, clearly marked "for professional use only." And this is wrong. Whether we can play or sing a single note, music enriches most of our lives—it entertains us, comforts us, and inspires us. The ideas of contemporary science, even and especially a field as seemingly remote from most of our everyday lives as cosmology, can do the same.

I recommend starting with the color illustrations at the center of the book—enjoy them on purely aesthetic grounds before diving into the text. Then come back to them at the end—and in the midst—of your reading. I'm hoping that you will find, as I do, that a deeper understanding of the science does not detract from the ability to appreciate the beauty of nature, but adds a richness to the experience—an ability to sense the nuances and undertones that lie beneath the surface. And I also hope that you will emerge from a first viewing with questions—cosmology is definitely a curiosity-driven sport.

No prior knowledge of science of any kind is assumed. Readers who have enjoyed some of the recent excellent books devoted to cosmology or particle physics should find that the first three chapters cover familiar ground, but even if the last science book you opened was an eleventh-grade textbook, you should be fine. (And in that case I'm pretty confident that this book will be much better than your last one.)

Which brings me to my first caveat: some of the concepts in this book (and any other on fundamental physics) are difficult to digest the first time through. This is not a reflection of your ability to understand and enjoy the topics—everyone, including every professional scientist, has had to struggle with them at some point as well. Wrestling with new ideas is at the heart of science (and a lot of fun)—the trick is not to lose confidence in yourself. I am certain that most readers will grasp the main concepts in this book long before I master a decent overhead serve on the tennis court.

However, this book is not intended as a course in basic physics, but as a means of transporting the reader to the brink of the most exciting frontier in science today—the quest to understand the composition of the Universe. Using a concrete and very visual implementation of Einstein's theories of space and time, we are making amazing progress in the hunt for the dark energy and dark matter that comprise the bulk of the cosmos. I hope you will join us.

Evalyn Gates
Chicago, Illinois

Acknowledgments

I am deeply appreciative of the help, support, and encouragement of many people during the past few years.

My agent, Lisa Adams, has been fantastic from the beginning—her encouragement and expert advice were invaluable. It was also a pleasure to work with my editor, Maria Guarnaschelli, whose guidance and direction were essential in creating the final version of the manuscript. Her assistants Margaret Maloney and Melanie Tortoroli somehow managed to direct all of the pieces where they needed to go.

The book was greatly improved in many ways by the comments, suggestions, and gentle critiquing (from a safe distance) of my readers. I would especially like to thank Martha Bohrer for her patient and careful reading from a nonscientist's perspective—her insights and questions were extremely helpful in clarifying the presentation of difficult concepts. I owe a special debt to several of my colleagues who generously agreed to review various sections of the book: David Bennett, Robert Caldwell, Sean Carroll, Scott Dodelson, Lucy Fortson, Josh Frieman, and Cole Miller. In addition, I would like to thank Vikram Dwarkadas, Mike Gladders, Andy Gould, Geza Gyuk, Bhuvnesh Jain, John Stachel, Albert Stebbins, John Terning, and Kip Thorne for discussions on certain topics; also Marlies Carruth and Don Santoski for their helpful comments. Fabian Schmidt, a graduate student at the University of Chicago, was kind enough to help with the translation of Einstein's notes from the original German. Marv Bolt,

Vice President for Collections at the Adler Planetarium, gave expert advice regarding many of the historical references in the book.

The illustrations are the work of a talented artist, scientist, and friend—José Francisco Salgado. It was a joy to work with him in creating the many figures in the book, and I stand in awe of his ability to turn my ideas and vague sketches into beautiful and descriptive artwork. For help in locating the right image or permission for its use, I thank Charles Alcock, Chuck Bennett, Timothy Carnahan, Stephane Colombi, Kem Cook, Pedro Duque, Roni Grosz, Jacqueline Hewitt, William Keel, John Ruhl, Volker Springel, John Stoke, and Tony Tyson.

I am particularly grateful for the ongoing support and encouragement from friends, including Martha Bohrer, Marlies Carruth, and Lucy Fortson, who were there at the beginning and who have helped in so many ways.

Finally, a special thanks to my family members for their love and support, especially my husband, Eric, who agreed not to read the book until it was finished (except for one chapter, which was greatly improved as a result of his carefully phrased comments and questions), and my parents, who encouraged me to follow my curiosity. To Greg, who asked about colliding black holes over 20 years ago; to Kendall, who graciously attended a special lecture about black holes on her birthday; and to Jordan, who may yet explore other universes (and who said it would never be finished)—this book is for you.

Glossary of Acronyms

Physicists can't seem to resist creating acronyms or abbreviations for particles, experiments, and theories. Below is a brief list of these terms (and a few useful definitions) that are used throughout the book.

CDM—*cold dark matter*. The generic name for the type of dark matter that best fits the current data. WIMPs are cold dark matter; so are axions.

CDMS—Cryogenic Dark Matter Search

CMB—*cosmic microwave background*. The CMB is light from the early Universe, often referred to as the "afterglow of the Big Bang."

EROS—Expérience pour la Recherche d'Objets Sombres ("Dark Object Research Experiment" in English)

Halo (or dark matter halo)—a (roughly spherical) cloud of dark matter. The visible components of a galaxy or a cluster of galaxies are embedded in much larger dark matter halos (Chapter 9).

HST—Hubble Space Telescope

JDEM—Joint Dark Energy Mission

Λ—*Lambda*. This symbol is used for the (as yet unknown) dark energy component of the Universe.

ΛCDM—*Lambda Cold Dark Matter*. ΛCDM is the currently favored cosmological model, in which dark energy is the main component of the Universe, and most of the matter is in the form of cold dark matter.

LHC—Large Hadron Collider

LMC—*Large Magellanic Cloud*, a nearby galaxy.

LSST—Large Synoptic Survey Telescope

MACHO—*massive astrophysical compact halo object*. MACHOs are dark objects such as faint stars, white dwarfs, or planets in the galactic halo, and are composed of normal matter.

MACHO collaboration—one of the teams searching for MACHOs.

Main components of the Universe:

- **Normal Matter (5%)**—matter composed of the quarks and electrons that are part of the Standard Model of Particle Physics (Illustration 1.1 on page 10).
- **Dark Matter (23%)**—a new kind of matter that can't be seen directly.
- **Dark Energy (72%)**—the unknown substance that is fueling the accelerated expansion of the Universe.

Nucleosynthesis—the formation of the lightest elements (the nuclei of deuterium, helium, and lithium) that occurred about one minute after the Big Bang.

OGLE—Optical Gravitational Lensing Experiment

SDSS—Sloan Digital Sky Survey

SKA—Square Kilometer Array

SMC—*Small Magellanic Cloud*, a nearby galaxy.

w—the "w" parameter is used to characterize dark energy.

WIMP—*weakly interacting massive particle*. WIMPs are one of the leading candidates for cold dark matter.

VCR—*video cassette recorder* (the author recognizes that this term may quickly disappear from common usage).

EINSTEIN'S TELESCOPE

CHAPTER 1

What Is the Universe Made Of?

We thought we were so close. The grand overall picture of the Universe was nearly in place, and scientists were busy finalizing the details of its structure and completing the mosaic of its history. Armed with bigger and better telescopes, astrophysicists settled in to appreciate the unprecedented clarity of their view of the cosmos, only to find that their new instruments revealed a Universe that didn't act at all the way it was supposed to. A glorious cosmic census of the hundreds of billions of galaxies, each shining brilliantly with the light of hundreds of billions of stars, should have allowed us to sum up all the matter in the Universe—a key ingredient in determining its ultimate fate. Instead, our observations insist that the stars and galaxies we can see are only a tiny fraction of all the matter that exists in the Universe—the rest is invisible to our telescopes, composed of new kinds of particles called *dark matter* that are far more exotic than anything we have ever seen or created in even our most sophisticated laboratories.

Even more outrageous, however, is the possibility that most of what exists in the Universe is not matter of any kind, but some strange new energy that is fueling the accelerated expansion of the Universe. An expanding Universe is one of the wildest predictions of Einstein's theory of general relativity. Space itself has been expanding since the earliest moments of time, and the distant galaxies that dot the heavens appear to be rushing away from one another as the space between them increases. But cosmologists expected gravity—which pulls mass together—to oppose this cosmic stretching and slow the overall rate of the expansion. What we have found is exactly the opposite, and it has pulled the ground out from under our understanding of the cosmos. It is as if, on your drive home from work tonight when you expect gravity to keep your car firmly planted on the solid pavement of the highway, you were to suddenly find yourself airborne, your car soaring high above familiar ground—and headed even higher. The expansion of the Universe is not slowing down, but speeding up. The Universe is expanding ever faster toward an unknown future, powered by something that has been dubbed *dark energy*, whose existence was unexpected and whose presence remains unexplained. Gravity, once thought to be the dominant player on the grand scales of the Universe, is ceding control of the cosmos to this unknown and unseen substance.

In a sense, science has fallen through a rabbit hole, and the world in which we find ourselves is far more preposterous than any Carrollian adventure. We have been handed a Universe that is overwhelmingly dark to our eyes and our telescopes—one that is roughly three parts dark energy to one part dark matter, with only a pinch of the familiar sprinkled throughout the cosmos like a handful of glitter on a vast sea of dark felt. The well-traveled road of twentieth-century physics is still there far beneath us, but we have been launched into a new dimension of exploration.

Such a radical new worldview demands an equally radical new means of observation—and Einstein's theory of general relativity, which initiated the last major revolution in our understanding of the Universe, is leading the way to the next revolution by making it possible for scientists to use space itself as a telescope. Forged by gravity, the lenses in this cosmic telescope are far more powerful than anything we could hope to build here on Earth.

According to Einstein, space and time are warped by the presence of mass of any kind. Every planet, star, and galaxy creates a dimple in space—the more massive the object, the larger the dimple. Dark matter, whatever it is and wherever it may congregate, also deforms the space around it. Light traveling through the Universe follows the curves of these dimples and bends around a massive object exactly as if a giant lens had been placed out in space. The warps in the fabric of space act as gravitational lenses to bend and deflect light in the same way that light is deflected by less exotic lenses fashioned out of glass or plastic.

This very practical application of what may appear to be a rather abstract theory is, in fact, one of the most powerful tools available to the current generation of cosmic explorers. Just as our understanding of quantum mechanics led to an innovative imaging technology—MRI—that allows doctors to view the inside of the human body in ways that could not have been imagined 100 years ago, so too has general relativity given astronomers a new means of imaging the cosmos. Einstein's theory is enabling us to push the limits of our vision to new extremes, revealing previously unseen details of the dark components of the Universe. With the aid of gravitational lenses, we have been able to magnify into view the most distant galaxies; to find otherwise undetectable planets far outside the Solar System; and to discover black holes as they whirl through space.

More important, we can use gravitational lenses in the search for dark matter and dark energy. We have already begun to trace out the dark matter that enshrouds galaxies and dominates the mass component of the cosmos, and we are starting to delve into the structure of the Universe in regions where no light can be found. Gravitational lensing is the only way to map the strands and filaments of dark matter that stretch across the vast reaches of the Universe in a web that was sculpted by gravity and stamped with the imprint of dark energy. The vistas we uncover with this new gravitational telescope—*Einstein's Telescope*—will take us further than ever before in deciphering the riddles of dark matter and dark energy, providing answers that may unlock the door into a deeper understanding of the fundamental nature of space, time, matter, and energy.

THE BIG QUESTION

Before we can look for answers, we first have to understand the question. As we enter the twenty-first century, the most urgent and compelling question facing scientists who study the Universe is also one of the oldest questions posed by humans in their attempt to understand the world around them:

What is the Universe made of?

This question has been the focus of scientific endeavors for millennia, and there is no doubt that we have made amazing progress. We now understand the basic nature of the matter that makes up everything we can see, from the Earth and everything on it to the stars in the most distant galaxy. At the most fundamental level, they are all composed of the same quarks and electrons that form the atoms and molecules in this page, and whose properties have been mapped out in great detail over the past 100 years or so. Toward the end of the twentieth century, some (misguided) pundits were even willing to claim that we had the answer to this ancient question—that the end of science was in sight.

They were spectacularly wrong.

The rush of scientific adrenaline that is reenergizing the study of this question today is fueled by a new understanding of the question itself, a recognition that the boundaries of our previous searches were far too limited; that they focused on but a tiny fraction of the bizarre constituents of our Universe. Data from opposite ends of the scientific spectrum, from recent experiments probing the subatomic world to new observations of the distant cosmos, have joined forces to reveal an incredible picture of the Universe—one in which the question *What is the Universe made of?* has been completely rewritten. We now know that most of the matter in the Universe is very different from the matter we have so thoroughly dissected here on Earth, and that most of the Universe is not even in matter of any kind, but in some new substance whose strange properties we do not understand at all.

From a scientist's point of view, this new state of affairs is fantastic. These new data have confirmed many of the predictions of the Big Bang theory—our current model of the Universe—and strengthened the case for its audacious description of the Universe and its history.

And at the same time, these data have handed us a new mystery: just what are these dark substances that dominate the cosmos?

The quest to understand what things are made of—which also helps us to determine where we fit into the grand scheme of things—is one of the most essential, most human of endeavors. Our very survival, as individuals and as a species, has depended on this fundamental curiosity, which seems to be hardwired into us from birth. We begin to explore our world with whatever senses are available to us, expanding our very definition of the world as fast as we are able to reach out and grasp any tiny sliver of the next realm. This innate curiosity has taken us far beyond our own backyard, motivated us to develop ever more sophisticated tools that can extend our senses far beyond those we are born with, and encouraged us to combine our critical thinking skills with great leaps of imagination in order to make sense of what we perceive with these new tools. The most exciting moments in these adventures come at the point when we realize that we are about to enter a new regime—when we have just landed on the shores of a new continent, touched down on a new planet, or plunged to new depths beneath the surface of the ocean; the point at which we begin to grasp the enormity of what it is that we *don't* know, and begin to chart the exploration of these foreign territories.

THE STATE OF THE UNIVERSE

Scientists have made astonishing progress in understanding the extreme realms of the Universe—the domains of the smallest and the largest. At the tiniest scales of subatomic particles, the Standard Model of particle physics has given us a solid, and so far unshakable, framework of the basic building blocks of normal matter. And at the other end of the spectrum, general relativity, which forms the foundation of the Big Bang model of the Universe, has also proven to be robustly successful.

The description of the microcosmos—the subatomic world—defines the smallest known particles of matter and the ways in which they interact with each other. What are the ultimate constituents of an atom, for example, and what holds them together?

On the experimental side, there are two main ways to explore the particle world. The first approach involves taking things apart. Physicists have dissected and classified everything that they've been able to get their figurative hands on. And they haven't been particularly delicate about it. The current state of the art basically consists of smashing two bits of matter together at incredibly high energies and watching to see what emerges—if only for the briefest of moments—from the collision. This method may sound about as sophisticated as a 10-year-old's hammering at a VCR to disassemble it, but thanks to Einstein's famous $E = mc^2$ equation, some of the energy of the collision can be transformed into new particles. The more energy put into the collision (the faster the initial particles are speeding toward each other), the heavier (and rarer) the particles that can be produced as a result. So there's a bit of twist here: breaking particles into the smallest possible pieces requires high-speed collisions, but as a bonus this process also allows us to create and search for new, and heavier, particles.

The second method involves setting a trap and waiting to see what particles are captured or even just pass through, leaving behind a trail of evidence from which we can deduce their existence. To build such a trap we need a target—something the particle can bump into, interact with, or scatter off of. For particle hunters, targets come in a wide variety of sizes and composition: a small crystal of silicon; a tank containing 50,000 tons of water; or even the upper atmosphere of the Earth. Detectors attached to these targets then collect and count up the end products of the interaction, noting the presence of charged particles, recording brief flashes of light, or sensing a minute increase in temperature of the target.

During the last century both methods were put to work in a host of innovative experiments. And while the experimentalists were busy happily collecting all sorts of strange bits of matter, their theoretical counterparts, newly armed with quantum mechanics, were constructing mathematical models that could incorporate these bits into a coherent overall description of the subatomic world. Such models often led to new predictions, which were then tested by the next generation of experiments. New particles were postulated—and found.

Strange interactions were observed—and explained by a few basic rules. The picture that finally emerged, known as the *Standard Model of particle physics*, is simple, elegant and, most important, consistent with all of the data. It describes the fundamental particles that constitute *normal matter*—the matter that makes up the Earth and everything on it, and the only kind of matter that we have ever actually detected—and the ways in which they interact with one another.

Pick up anything near at hand. A book, your dog, the air you're breathing—all are made from the same essential tiny bits of matter. Whatever it is, it can be broken down into molecules of various kinds, each molecule composed of one or more atoms. The different kinds of atoms—hydrogen, oxygen, and gold, for example—are laid out in the familiar table of the chemical elements, which can be found on the wall of any high school chemistry classroom. Each of these atoms consists of a nucleus enshrouded by a cloud of electrons. The nucleus can be further broken down into its constituent neutrons (which have zero electric charge) and protons (which are positively charged). Neutrons and protons are in turn composed of three *quarks* each. So far, quarks have not yielded to further dissection—they are considered the smallest building blocks of protons and neutrons.

Travel now to the surface of the Moon to scoop up a handful of dust or to the center of the Sun, assuming you could survive such a trip, to sample the 15-million-degree hot gas. Zoom out to any star in the galaxy or burrow down to the center of the Earth. Visit one of the major particle physics centers, such as Fermi National Accelerator Laboratory (known to its friends as Fermilab) outside of Chicago, and collect whatever comes out of one of their experiments. All of this matter—regardless of how hot or cold, rough or smooth, abundant or extremely rare—is made of the same basic stuff: the same quarks and electrons that make up the atoms in your body or your book; heavier but otherwise identical copies of these quarks and electrons; and ghostly particles called *neutrinos*[1] that are produced in nuclear reactions.

The list of particles in the Standard Model includes six kinds of quarks (*up* and *down*, *charm* and *strange*, *top* and *bottom*); the electron, the *muon*, and the *tau particle*; and three flavors of neutrinos, known as the electron neutrino, the muon neutrino, and the tau neutrino. These particles are grouped into three "families,"[2] and the complete catalog of normal particles is shown in Illustration 1.1a. One way or another, all of these particles have been detected, and various combinations of

Normal-Matter Particles

Family I	Family II	Family III
u up quark	**c** charm quark	**t** top quark
d down quark	**s** strange quark	**b** bottom quark
e electron	**μ** muon	**τ** tau
ν_e electron neutrino	ν_μ muon neutrino	ν_τ tau neutrino

Force Carriers	Force
γ photon	Electromagnetism
W^+ W^- Z W and Z particles	Weak nuclear force
g gluon	Strong nuclear force
? graviton	Gravity

ILLUSTRATION 1.1 The Standard Model of Particle Physics. (a) The fundamental building blocks of all normal matter and the force particles that mediate interactions between them.

them produce everything else made of normal matter. For example, to construct any element in the periodic table, all you need are a few quarks (two up quarks + one down quark = proton; two down quarks + one up quark = neutron) and some electrons.

Further, each particle has an *antiparticle*—a copy of the particle that is identical to it in every way except that the copy carries the opposite electric charge. A particle known as the *positron* has the same mass as the electron but a positive charge. Each of the quarks has its antiquark twin, and an *antiproton* is formed out of two anti-up quarks and one anti-down quark. These particles are collectively known as *antimatter*, but they have normal gravitational interactions and are considered part of the Standard Model.

Antimatter may sound like something out of science fiction—and it is true that if a proton and an antiproton meet they will annihilate each other—but it's not really all that exotic. We are still trying to understand why the Universe is composed almost entirely of matter, not antimatter, but antimatter particles are produced both by nature (in the decay of some nuclei, for example) and by scientists (beams of antiprotons are created at Fermilab and positron beams are created at the Stanford Linear Accelerator Center—SLAC—in Menlo Park, California) on a regular basis. Either way, the antimatter doesn't stick around very long. As soon as it finds a bit of matter, it disappears in an act of mutual annihilation, the mass of both particles transformed—at least momentarily—into pure energy.

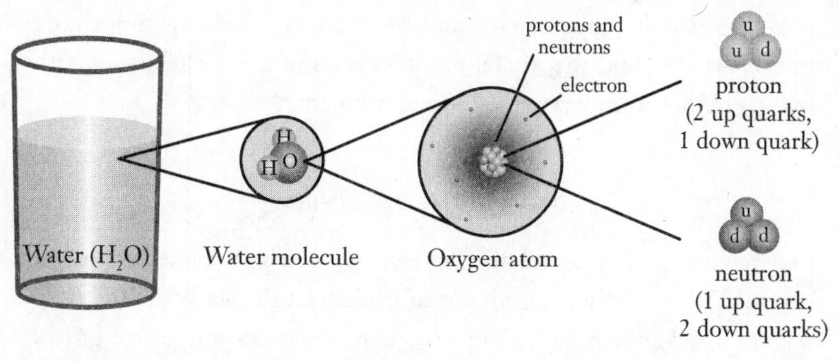

(b) A glass of water broken down into its most basic components.

The interactions of all of these particles—both matter and antimatter—are governed by four basic forces, as listed in Illustration 1.1a. Gravity is probably the most familiar (and at the same time possibly the least understood). Electricity and magnetism, which appear to be two different things, are actually just different manifestations of the same electromagnetic force. Less familiar are the strong and weak nuclear forces. Each of the forces has a force particle associated with it, and matter particles interact with one another through the exchange of these force particles. Charged particles, such as the electron, interact with other charged particles via the *photon*, the carrier of the electromagnetic force. Quarks are bound together in the nucleus by *gluons*, which transmit the strong nuclear force. And the weak-force *W* and *Z particles* (also known as *W* and *Z bosons*) mediate interactions between electrons and neutrinos, for example. There is probably also a *graviton*—a particle associated with the gravitational force—but we haven't yet observed one. This is one of the key questions remaining in particle physics.

The Standard Model of particle physics is a mathematical description of the subatomic world—the fundamental particles that comprise normal matter and their interactions—that has been extremely successful. However, the Standard Model does not explain everything, and serious questions remain—Why are some particles heavier than others? Why are there three families of particles instead of just one? Is there a quantum theory of gravity?—but so far we haven't been able to find any weak points in the Standard Model itself. Physicists have poked, prodded, ripped apart, and smashed together this model and these particles, and the model has passed all of these challenges with flying colors—a truly remarkable achievement.

BACK TO THE BEGINNING

Our picture of the cosmos also achieved stunning successes in the last century. We now have mountains of evidence to support the *Big Bang theory*, which describes a Universe that began as hot dense gas of energy, then cooled over billions of years in an expansion of space itself

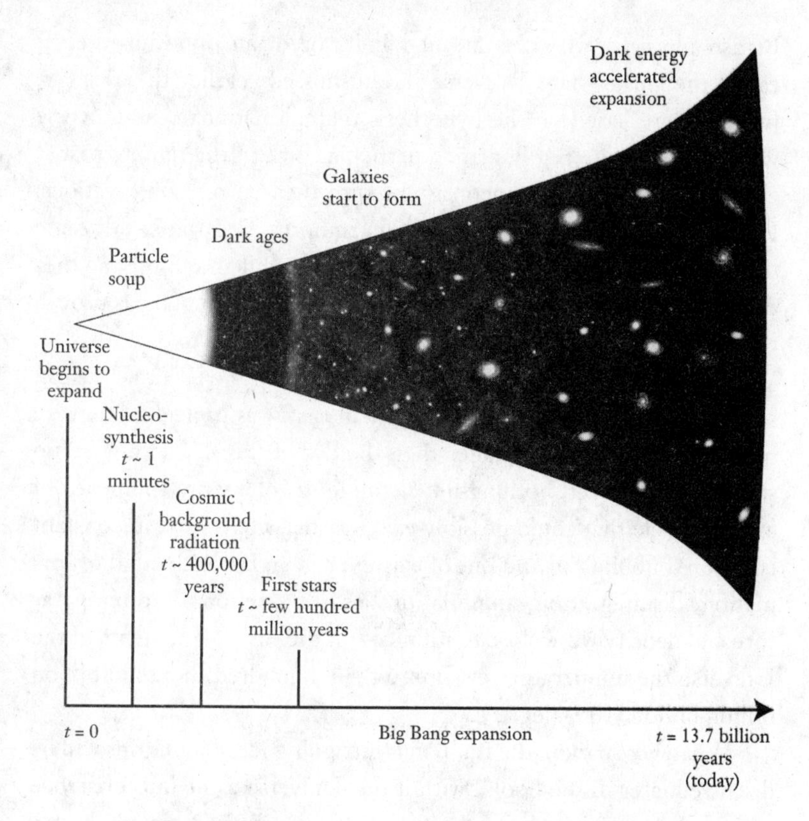

ILLUSTRATION I.2 The history of the Universe. This diagram highlights the key eras in the history of the Universe as it has expanded and cooled over the past 13.7 billion years.

that continues even today. The Big Bang, in spite of what its name implies, was not an explosion in space, but rather the beginning of space and time—or, more precisely, as close as our current understanding of physics can take us to this starting point.

The term *Big Bang* is often used in two ways: the *Big Bang theory*, which refers to the model of a Universe that expanded from an initial hot, dense, state; and the *Big Bang* itself, which denotes an infinitely dense, infinitely hot starting point of the Universe (technically known as a singularity), which we are still trying to understand. *Big Bang* was actually first used in a 1949 BBC radio program by Fred Hoyle, a

British physicist who was arguing in favor of an opposing theory, called the steady-state Universe. He dismissed "earlier theories . . . [which] were based on the hypothesis that all matter in the universe was created in one Big Bang at a particular time in the remote past,"[3] claiming that they were inconsistent with then-current observations. However, Hoyle was in the scientific minority and the steady-state theory is currently not supported by the data, while the Big Bang theory (the term was subsequently adopted by its proponents) has proven to be amazingly consistent with observations and data to date.

About 13.7 billion years ago, the part of the Universe that we can see today, and probably much more than that, was packed into a very hot, very dense region smaller than the dot of this *i*. There were no stars or galaxies yet, nothing but a roiling sea of energy composed of every particle that could possibly exist; particles that were in constant collision, bubbling in and out of existence in an endless round of formation, disintegration, annihilation, and production. The temperature and density were close to infinity—in the earliest moments of the Universe the temperature was more than a hundred thousand billion billion billion (10^{32}) degrees.

Whatever sparked the Big Bang (intriguing ideas about this will be discussed later in the book), with it our Universe came into existence and space began its expansion. And as it expanded, it cooled. By the time the Universe was one-millionth of a second old, the temperature had decreased to a mere 10 trillion (10^{13}) degrees, heralding a new era in cosmic history. Particles—protons and neutrons—began to condense out of the cosmic soup.

Until this milestone temperature was reached, the Universe had been too hot for composite particles to be formed and stick around for any length of time—as soon as three quarks came together to form a proton, some other particle would crash into the proton with more than enough energy to blast it apart again. Imagine a small room full of energetic preschoolers allowed to run wild. Towers of blocks might be built, but they would just as quickly be knocked apart. Only as the energy of the youngsters began to flag, and they stopped careening from one end of the room to the other, would towers endure long enough to be admired by a parent at the end of the day. Similarly, protons and neutrons could populate the cosmos only when the typical energy of particles (which decreases with the temperature) dipped below the energy level required to disintegrate these particles back into their constituent quarks.

At about one minute after the Big Bang, the temperature dropped below 1 billion degrees, and protons and neutrons themselves became bound together to form the lightest of the elements in a process known as *nucleosynthesis*. All of the heavier elements were formed much later, forged in stars and stellar explosions, but these light elements—mainly hydrogen, with a sprinkling of helium and a hint of lithium—constituted the first wisps of gas that permeated the Universe, gas that would eventually form the first stars.

The conditions were not yet ripe for the creation of stars, however. The earliest Universe was very smooth, with only the tiniest of lumps in the sea of particles and energy. These lumps represented the seeds of stars and galaxies, seeds that would eventually grow under the relentless action of gravity—once it won its tug-of-war with the expansion of the Universe. In order for this to happen, matter had to gain the upper hand.

For tens of thousands of years after the Big Bang, the energy content of the Universe was mainly in the form of radiation—light and fast-moving particles. Matter (in the form of slow-moving particles) made only a minor contribution to the energy budget. Fortunately for us, the ratio of matter to radiation changed as the Universe expanded, increasing until matter achieved majority status and radiation became less and less important. This change is essential to our existence. In a Universe filled with radiation, the expansion of space overpowers the efforts of gravity to pull more matter into overdense regions, and the small seeds of future stars and galaxies cannot grow. Matter must become the dominant component of the cosmos before structures can begin to form—structures that include our Galaxy.

In the very beginning, the cosmic energy budget was almost entirely given over to radiation. Within the primordial soup were particles of light (known as *photons*) and particles of matter, but even the matter particles that were present were counted as radiation because particles at very high energies (temperatures) are relativistic—they move at nearly the speed of light. The energy of their motion dominates their total energy, and they behave more like radiation (light) than like matter (something with mass).

As the Universe expanded and the temperature dropped, the energy of the matter particles also decreased until they slowed enough to become nonrelativistic (the energy in the motion of the particle becomes small relative to the energy in its mass). Slow-moving matter particles effectively switch teams and are counted as matter instead of radiation. However, even after most of the matter particles became nonrelativistic, radiation continued to dominate the cosmos—there was still more energy in the remaining radiation (mainly light and neutrinos) than in matter.

The expansion of the Universe would change this. Consider a chunk of space, filled with particles of matter and light. As the Universe expands, the volume of this chunk increases, but the number of particles remains the same—and thus their density decreases. This dilution applies to both matter particles and light particles. The light particles pay an extra penalty, however. As space stretches, light waves are also stretched—the wavelength of light becomes longer, and since the energy of light decreases as its wavelength increases, the energy density in light is reduced further by this stretching factor. Thus, as time goes on and the Universe expands, the energy density in light decreases faster than the energy density in matter, and eventually the energy in matter wins out.

Roughly 50,000 years after the Big Bang, this key transition took place. Matter for the first time was on equal footing with radiation, and the development of structures in the Universe began. As the Universe passed the 56,000-year mark and the temperature dropped below 10,000 degrees, the tiny lumps in the density distribution—a few more particles in one place, a few less in another—began to grow under the influence of gravity.

At this temperature it was still too hot for atoms. The nuclei of the light elements that had formed earlier were unable to hold on to the electrons that whizzed past them. The positive charges of the naked nuclei and the negative charges of the free electrons made it very difficult for light particles—photons—to travel through the soup of particles. Light particles interact (through the electromagnetic force) with any charged particle they come close to, and these early particles of light careened from one charged particle to another, changing energy and direction after each collision. The net effect was the primordial equivalent of a thick fog, impossible to see through.

When the temperature dropped below 3,000 degrees, about 380,000 years after the Big Bang, electrons and protons finally came together to form neutral atoms—and the light was freed. With essentially no charged particles left to get in the way, the light particles created in the Big Bang could travel unimpeded through space and time. They continue to stream through the Universe almost 14 billion years later. Known as the *cosmic microwave background*, or *CMB*, this light is the afterglow of the Big Bang—a remnant of the early Universe that has proven to be a gold mine of cosmological information. When we detect this light today, we are seeing the outline of the primordial fog—an image of the Universe as it appeared less than 400,000 years after the Big Bang.

Paradoxically, the emancipation of the ancient light from the Big Bang also heralded the beginning of the cosmic dark ages—a period when almost no new light was created. During the dark ages stars did not yet exist—leaving us bereft of information or images of millions of years of cosmic history. In the darkness, gravity continued to exert its inexorable pull, and the lumps in the distribution of matter grew ever larger as the gravitational force of each lump attracted more and more matter.

The dark ages ended with the creation of the first stars. We don't yet know the precise moment when the first star was born, but it was roughly a few hundred million years after the Big Bang. Clouds of hydrogen gas collapsed under their own weight, and the resulting increase in temperature and pressure in the core of these clouds ignited the nuclear processes that lit up the star—and the cosmos. The formation of the largest of these new stars was fast and violent, and the end was even more so. After a brief existence, they exploded in a brilliant release of energy and particles—and in the process produced heavier elements such as iron that are essential for our existence.

Over the next 13 billion years, small galaxies of stars formed, often growing larger through a process of mergers and acquisitions, and groups of galaxies came together in clusters containing hundreds or even thousands of galaxies. As new stars continued to light up in these galaxies, the disks of debris that accompany stellar formation sometimes coalesced into planets—giant gas planets, small icy planets, and mineral-rich rocky planets. Our own star, the Sun, saw first light about 5 billion years ago, and the Earth and its companions in the solar system were formed from the disk of gas and dusk that swirled about it.

And the Universe continued to expand and cool. Today we look out at a Universe that is 13.7 billion years old, filled with over 100 billion galaxies, each containing hundreds of billions of stars. The temperature has fallen to almost zero. The CMB, which acts as a cosmic thermometer, initially glowed at a temperature of 3,000 degrees, but it is now just 2.73 degrees above absolute zero. And it's likely to get even colder in the future.

The evidence for this description of the history of our Universe has three key components. We will cover each of these in depth later in the book, but to state them briefly, we have observed directly

1. *The expansion of the Universe.* In 1929, Edwin Hubble published the first paper presenting evidence that distant galaxies are all moving away from us, indicating that we live in an expanding Universe.[4] Subsequent observations up to the present have confirmed this result and measured the rate of the expansion (with a precision of about 5%).[5] This expansion implies that if we were to run a video of the Universe in reverse, we would see the galaxies getting closer and closer as we went back in time. The Universe would get hotter and denser until it eventually reached of point of nearly infinite temperature and density—the Big Bang.

2. *The formation of the light elements.* Nucleosynthesis makes precise predictions for the abundance of these light elements in the early Universe. As astronomers have increased the precision of their observations of primordial helium, deuterium, and lithium, all of which were produced in vastly different amounts, their results continue to be in outstanding agreement with predictions of the Big Bang model.[6]

3. *Cosmic microwave background.* The CMB was first detected in 1965,[7] and its temperature was consistent with predictions for a Universe that began in a hot, dense state and then expanded and cooled for 10 billion to 15 billion years. This observation effectively nailed the case for the Big Bang model.

The Big Bang also makes another prediction for the CMB. The early Universe was almost a perfectly smooth fluid, except for some extremely tiny, but extremely important, bumps and dips. In some places the fluid was slightly denser, in others slightly less concentrated. At the time the CMB was generated, these imperfections were on the order of one part in 100,000—the merest suggestions of a departure from uniformity. But, in the gravitational equivalent of the rich getting richer and the poor poorer, the small overdensities grew ever larger over billions of years as their gravity was able to attract additional matter. From these tiny initial seeds emerged the structure we see in the Universe today—galaxies and clusters of galaxies strung across the Universe on a vast web of matter.

The CMB, which is an image of this early fluid (the cosmic fog), should also exhibit the same tiny inhomogeneities. In 1992, the *COBE* (*Cosmic Background Explorer*) satellite made a very precise all-sky map of the CMB. At every single point on the sky COBE measured almost exactly the same temperature of 2.73 degrees Kelvin—but not quite. Seen in this first map of the CMB were tiny temperature variations at the level of about one part in 100,000—just as expected. The observed pattern of hot and cold spots reveals the imprint of the early lumps in

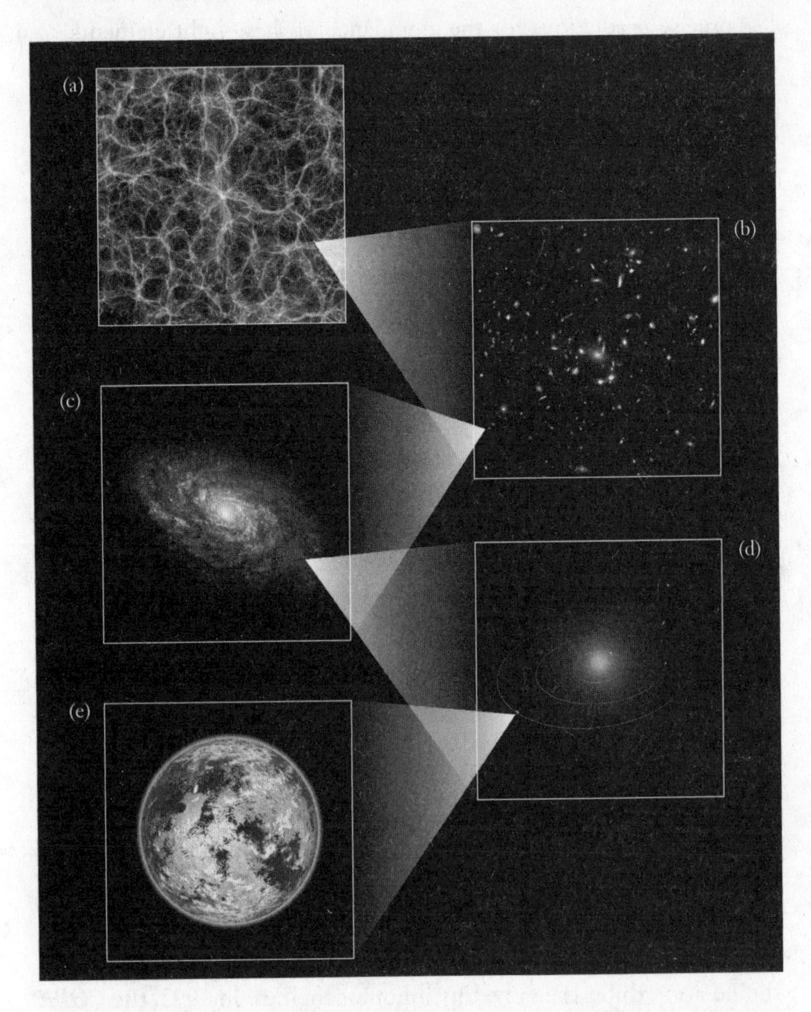

ILLUSTRATION 1.3 The organizational chart of the Universe. The Universe has structure on a vast range of scales. (a) This computer simulation of the cosmic web of dark matter represents a section of the Universe roughly 1 billion light-years across. In the densest regions of this web are found groups or clusters of galaxies (b) that contain tens to hundreds or more galaxies and can span 10 or 20 million light-years. Galaxies come in a variety of shapes and sizes—shown here (c) is a typical spiral galaxy (NGC 4414) much like our own Milky Way, with a diameter of about 56,000 light-years. Each galaxy contains billions of stars (d), some of which are orbited by one or more planets (e).

the cosmic fluid. (The most recent microwave map of the sky is shown in Color Illustration 10).

In understanding why scientists are so strongly convinced that the Big Bang model is an excellent (and at present the best) description of the evolution of the Universe, it's important to note that each of these three key observations comes not from a single experiment or telescope, but from many different scientists using different methods and different instruments. Further, each of these three observational pillars of the Big Bang model is looking at a different component of the Universe and at a different point in time in the history of the Universe. The expansion rate is determined by observation of galaxies and the stars in them at relatively recent times (for the most part within the past few billion years); the CMB observations collect light that was given off by the Universe itself billions of years earlier; while nucleosynthesis probes even further back in time, searching clouds of primordial gas for nuclei formed when the Universe was only a few minutes old. That all of these observations agree with the relatively simple (if definitely mind-blowing) Big Bang model is a true scientific tour de force.

WHAT WE DON'T KNOW

As our picture of the Universe and its contents has become clearer, our observations have also insisted on three other key points. First, there is much more matter in the Universe than is seen in stars and galaxies; second, most of this unseen matter cannot be composed of the Standard Model particles—the quarks and electrons and their heavier cousins described earlier—that we have come to know so much about; and third, most of the Universe is not in matter of any kind.

In 1933, Fritz Zwicky published the first observational evidence that the Universe had a dark component.[8] His goal was basically to weigh the largest, heaviest object he could find—a nearby cluster of galaxies. He set his sights on the Coma cluster, which contains more than a thousand galaxies, all moving around inside of the cluster at incredible speeds, whirled through space by the gravitational pull of the mass contained in the cluster. Zwicky focused on eight of these galaxies and found that they were all moving much, much faster than expected.

Gravity, although the weakest of the four fundamental forces, is the main mover and shaper of the Universe. And an understanding of how gravity works is what allowed Zwicky to attempt such an ambitious project. By observing how things move in space and making the (reasonable) assumption that gravity is responsible for these motions, we can determine the mass that must be present in order to make sense of the motions. Every object in the Universe tugs on all the other objects around it in a very predictable way—the more mass an object has, the stronger its gravitational pull; the stronger the pull, the faster the (change in) motion. Further, an effective speed limit is imposed on objects bound together by gravity, and objects that exceed this limit will escape the gravitational pull of the system.

Zwicky found that the galaxies in the Coma cluster had speeds of about 2 million miles per hour, implying that the cluster contained 50 times more mass than could be accounted for by adding up all the mass in the galaxies that defined the cluster. Three years later Sinclair Smith found a similar result for the Virgo cluster;[9] and since then observations of hundreds of clusters have reinforced the conclusion that most of the mass in clusters of galaxies is unseen. In his paper, written in German, Zwicky suggested the possibility that his observations implied the existence of *dunkle Materie*—"dark matter."

In spite of the catchy name, Zwicky's dark matter results were regarded with a healthy amount of skepticism by the astronomical community, which more or less ignored them for the next 35 years or so. It wasn't until 1970 that a second, and even more convincing, vote for dark matter came from another set of observations on a different scale. The stars and gas clouds in an individual galaxy are in orbit about the center of the galaxy, and their motions should reflect the amount of mass in the galaxy. In 1970, Vera Rubin and Kent Ford peered into the Andromeda galaxy (also known as M31) and, by measuring the speeds at which gas clouds orbited about the galaxy, found that it also must contain huge amounts of dark matter.[10] Data on over a thousand galaxies, including our own, now insist that roughly 90% of the mass in galaxies is dark.

The possibility that this unseen dark matter might be simply normal matter in a form that we can't see very well, such as extremely faint stars, or diffuse gas spread throughout the cluster or galaxy, was taken seriously for many years. We know that such stars and gas exist, and that is it very difficult to detect them, especially at large distances. However, the success of Big Bang nucleosynthesis in describing the formation of the light elements also puts strong constraints on how much normal matter can exist. Theoretical cosmologists start with a soup of protons and neutrons, turn on the reaction rates for the different ways in which they can combine to form deuterium (heavy hydrogen), helium, and lithium nuclei, and determine how much of each is produced. The only free parameter is the total number of protons and neutrons in the primordial soup. A little more or less, and the predicted amounts of deuterium or lithium change drastically.

Basically, a neutron, which is a tiny bit heavier than a proton, has two choices. It can decay into a proton and an electron, or it can join forces with a proton to form deuterium. Deuterium then pairs up with another deuterium to form helium, and occasionally some of the helium combines with other nuclei, ultimately producing a tiny amount of lithium. The key determining factor is the rate at which all of the interactions occur, and that rate depends on how many neutrons and protons there are—the higher their density, for example, the more likely it is that a proton and neutron will meet up to form deuterium before the neutron has a chance to decay.

When the predictions are compared with actual measurements of the abundance of these light elements, a startling result emerges—only about 5% of the total matter and energy in the Universe can be in the form of normal matter.[11]

Dark matter is also not made of antimatter, such as antiprotons. Essentially all of the antimatter particles created in the Big Bang were annihilated by matter particles at very early times in the history of the Universe, so only matter particles remain today.

Carl Sagan once said that "extraordinary claims require extraordinary evidence,"[12] and the claim that most of the matter in the Universe is something entirely different from normal matter certainly qualifies as extraordinary. The evidence in support of the nucleosynthesis results came from stunning new observations of the CMB. The pattern that the CMB produces across the dome of the sky contains a wealth of information, and as more and more measurements have been made, with increasingly higher precision, the secrets of the Big Bang have leapt out of the images.

One of the most important new results revealed by these measurements is the overall shape of the Universe. According to Einstein's theory of general relativity, space is no longer the static backdrop of the Newtonian world but responds to the presence of matter and energy, warping and deforming its shape according to the distribution of mass. Space and time are themselves inextricably linked in an entity referred to as *spacetime*. The Sun produces a small dimple in spacetime, the galaxy creates a larger deformation, and all of the matter and energy in the Universe affects its overall shape. Too much, and the Universe curves back in on itself like the surface of a sphere. Too little, and it curves in the opposite sense, opening up into the three-dimensional equivalent of the surface of a saddle. Just right, and its overall geometry is flat.

In late December of 1998 a balloon soared high above the Antarctic plateau, a detector riding beneath it recording the most detailed map of the CMB at that time. Encoded within the pattern that it observed on a patch of the sky was the unmistakable signature of a flat Universe—firmly ruling out a Universe that contains only the matter that we can see in stars and galaxies.[13]

Further, the amount of normal matter also leaves a signature in the CMB itself. At the time the ancient light of the CMB was emerging, the tiny lumps in the Universe were generating ripples within the cosmic fluid as matter fell into denser regions or out of less dense regions. Suppose you dropped two identical marbles into two identical buckets—one containing water, another motor oil—and snapped a picture of the surface of each fluid at exactly the same time after releasing the marbles. Even if all you could see from each photo was the pattern of

ILLUSTRATION 1.4 The Boomerang launch. On December 29, 1998, the Boomerang experiment was launched for a 10-day trip above the Antarctic continent. The 1-million-cubic-meter (28-million-cubic-foot) balloon, seen here with Mount Erebus in the background, flew at an altitude of 120,000 feet in order to obtain precise measurements of the cosmic microwave background.

ripples on each surface, you could easily tell which fluid was water and which was oil. With higher-precision photography (and detailed knowledge about the various kinds of motor oil), you could probably even decipher the type of oil.

In a similar way, the CMB provides a snapshot of the ripples in the early Universe that reveals much about the cosmic fluid. In particular, the most recent measurements have determined that only 5% of the matter and energy in the Universe is normal matter. Further, these measurements insist that 23% is composed of some entirely new kind of matter, about which we know only that it behaves gravitationally like normal matter, and the remaining 72% is not matter of any kind at all.

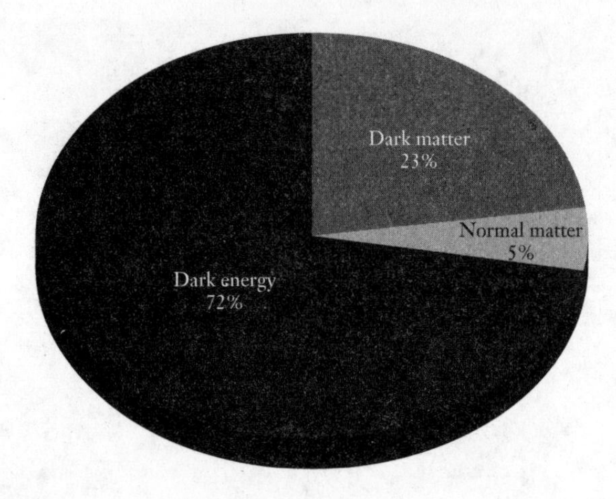

ILLUSTRATION 1.5 Cosmic Inventory: Our current understanding of the composition of the Universe.

Before moving on to the truly bizarre 72%, let's review the case for the 23% in a new type of matter. The combined evidence for it is overwhelming—we know that it's there because of its gravitational influence on the astronomical objects that we can see. The motions of stars within a galaxy cannot be explained unless there is an enormous amount of invisible matter in the galaxy whose gravity keeps the stars moving in their orbits. Galaxies themselves dance about each other at speeds that require the gravitational pull of far more matter than is seen in the combined mass of their stars and dust. And two other key pieces of evidence—the formation of the light elements in the early Universe and observations of the CMB—both insist that this extra mass must be some as yet unknown type of matter. Matter that is not seen, and in fact can't be seen directly by any of our telescopes. Dark matter.

Particle physicists have long suspected that there is more to the microcosmos than has yet ever been found. Although we understand the known particles and their interactions, we don't know the underlying physics that, at a deeper level, gives each particle its mass. Or why there are three copies of each particle, identical in every way except mass, such as the electron and its heavier counterparts, the muon and the tau particles. Or why the two great revolutions of the twentieth century—general relativity (which explains the large-scale behavior of the Universe) and quantum physics (which describes the small-scale interactions of the subatomic world)—are ultimately at odds with one another, resisting attempts to unite them in a quantum theory of gravity.[14] We suspect that our current model of the particle world is actually embedded in a larger model—one that encompasses the answers to at least some of these questions.

Most of these larger models also predict the existence of new particles with the right properties to be the invisible dark matter. So, although we still don't know what dark matter is, we can at least agree that it is matter of some kind (it has mass), with normal gravitational interactions, and that by extending our model of normal matter we can come up with some pretty good candidates.

The remaining 72% of the Universe is much more difficult to explain.

AN OUTRAGEOUS UNIVERSE

A critical component of the Big Bang theory is the concept that we live in an expanding Universe. Space itself is expanding. Distances between points in space are getting larger and distant galaxies are all rushing away from each other—the farther their separation, the faster they move apart. Incredible as this sounds, we have made many measurements of this expansion, beginning with the original observations of Edwin Hubble in 1929. Cosmologists have spent the past few decades determining with increasing precision the rate at which the Universe is expanding today, and in fact this was one of the key projects of the aptly named Hubble Space Telescope.

The data were converging about a reasonable number for the expansion rate, when two groups of astronomers decided to stir things up a bit. Instead of playing nicely in our own corner of the Universe, they began looking much farther afield, using distant stellar explosions (supernovae) to probe how this expansion rate has changed over the past few billion years. Supernovae are extremely bright, outshining an entire galaxy of stars for a brief cosmic moment (about a month) before fading away. Their brilliance allows us to see them at great distances, which is equivalent to looking back in time.

Telescopes collect light that was either emitted from an object in space (such as a star or a galaxy or the Sun) or reflected off of the surface of an object (such as the Moon or planets within our own Solar System). This light takes a certain amount of time to travel from the object to us: light reflected off the surface of the Moon takes about 1.3 seconds to reach the Earth, and light from the Sun takes a little over 8 minutes to get here. This means that when we look at the Moon, we are seeing it as it was 1.3 seconds ago, and the Sun's image is over 8 minutes out of date by the time it reaches us. The farther an object is from Earth, the longer it takes the light to arrive, and thus we are seeing the object not as it appears to an observer near it today, but as it appeared in the past. By looking far out into the distant Universe, we are able to see galaxies as they were billions of years ago. In 1987 a stellar explosion was detected in one of our closest galactic neighbors, the Large Magellanic Cloud (LMC). However, since the LMC is about 170,000 light-years away from us, the star itself actually exploded 170,000 years ago. It just took us this long to get the message.

The supernova hunters were looking for far more distant prey. And they were focusing on a particular kind of supernova that has another important characteristic—its intrinsic brightness is well-known. Because objects appear fainter as they get farther away, the distance to an object whose brightness is known can be deduced from how bright it appears to be. This relationship between distance and brightness makes these supernovae uniquely useful as distance markers, mileposts that can be seen billions of light-years from Earth. Accurate measurements of such distances are essential for tracing out the expansion history of the Universe.

About 15 years ago, two independent groups—the Supernova Cosmology Project, headed by Saul Perlmutter at Lawrence Berkeley National Laboratory, and the High-z Supernova Search Team, headed by Brian Schmidt of Australia's Mount Stromlo and Siding Spring observatories—began observing these distant supernovae, and what they found was completely unexpected.[15] Since every bit of mass in the Universe is attracted to all of the other bits of mass in it, their mutual gravitational attraction should be putting the brakes on the outward rush of the galaxies. But instead of slowing down, the supernova teams found that the Universe appears to be speeding up—our Universe is accelerating, flagrantly contradicting all previous predictions and implying the existence of a substance that challenges our most basic ideas of physics on both the largest and the smallest scales. This amazing announcement was acclaimed the "Breakthrough of the Year" by *Science* magazine in 1998, and it stunned the scientific community.[16] Something has to be fueling this acceleration—something very different from matter, dark or otherwise. Something that has been termed *dark energy*.

Dark energy is the major component of the Universe, constituting 72% of everything that is. We don't know what dark energy is, and we don't have any outstanding candidates either. Theorists are currently exploring a host of ideas and models that might explain dark energy—some wilder than others, but all involving radically new physics and none yet compelling enough to emerge as the main contender. Dark energy is, at this time, a complete mystery.

REDEFINING THE QUESTION

The history of science is punctuated by great revolutions in understanding that catapult us into previously unimagined landscapes. These revolutions can be sparked by new technology, new experimental results, or new ideas. Like Dorothy, whose entire worldview was upended as she stepped out of the black-and-white Kansas countryside into the Technicolor land of Oz, Antoni van Leeuwenhoek changed forever our concept of what is in a drop of water when he first looked through his microscope. The advent of quantum physics allowed physicists to tease apart atoms and their nuclei and opened up the subatomic world of quarks, electrons, and neutrinos. And at the opposite end of the size spectrum, the theory of general relativity gave us an entirely new description of the Universe on the grandest scales—one where space, time, matter, and energy are inextricably linked in a dynamic swirl of activity as unexpected as that seen by Van Leeuwenhoek through his microscope.

Our search to understand the composition of the Universe has led us to a new critical point. The question *What is the Universe made of?* has become, in essence, *What is dark matter?* and *What is dark energy?* If normal matter—the atoms and molecules that comprise everything we see—is only a tiny fraction of the total matter in the Universe, what is this dark matter that makes up the rest? What strange dark energy dominates our Universe and speeds its outward expansion in defiance of the gravitational pull of all of the matter within it? And how can we ever hope to answer these questions when neither dark matter nor dark energy can be seen with even our most powerful telescopes?

This new comprehension of the true nature of the question demands that our current focus on physics at the extremes of both the smallest and the largest scales be transformed into a partnership between those studying inner space and those exploring outer space. As astronomers search the far reaches of the cosmos for the imprint of how matter and energy have sculpted the Universe over the past 13.7 billion years, they are guided by the results and theories of particle physics. At the same time, astronomical observations are helping to focus the work of physicists as they peer down into the microcosmos, teasing out the hidden sectors of the subatomic world and postulating the existence of new particles and new dimensions.

Both pieces of the puzzle—the unimaginably large and the incredibly small—are essential for determining the composition of the Universe: If a new particle is proposed or detected, how does this affect the formation of galaxies or the rate at which stars burn their fuel? If galaxies are observed racing away from each other at ever-faster speeds, what does this tell us about the nature of "empty" space? If a snapshot of the Universe only a few hundred thousand years after the Big Bang reveals detailed images of ripples in the cosmic fluid of basic particles, can we reproduce those ripples with computer models containing the particles we know of—or do we need something dramatically different?

The study of the Universe is thus poised at an exciting and critical juncture. Although we know more than ever before about the answer to the ancient question—*What is the Universe made of?*—we also now know that there is far more to the question itself, and that what we don't know far outweighs what we do know. And we recognize that our current theories, our most sophisticated models, our most intricate mathematical calculations, may not lead to the answer. The answer to this question almost surely lies on the far side of the next great revolution in science.

CHAPTER 2

A Revolution in Space and Time

Long before the first beautiful images of the blue-green orb that we call home were sent back from space, the ancient Greeks arrived at the startling realization that the Earth is not flat but curved (and even measured its circumference).[1] We live on the surface of a giant sphere. Careful observations of the heavens eventually led to an even more drastic revision of our worldview. The other planets visible in the night sky had long held a special place in celestial observations, but the discovery that the Earth is but one of their number, in orbit with its sister planets about our home star, ejected us from what we assumed was our privileged position in the center of the Universe. And the boundaries of our Universe grew enormously when the stars, once seen as fixed points of light on a giant celestial sphere, were placed in their proper positions in a three-dimensional map of the Galaxy that extended much farther than ever before imagined.

At the beginning of the twentieth century, this model of a star-filled Universe described all that we thought existed. We had yet to discover that our Galaxy is only one of what we now know to be hundreds of billions of galaxies—"island universes"—in the observable Universe. And even more important, we had yet to incorporate time in our description of the Universe.

In 1915, Einstein completed the theory of general relativity, finalizing his description of a Universe in which space and time are not a mute and unresponsive stage upon which all action takes place, but

ILLUSTRATION 2.1 *Almagestum novum*, Giovanni Battista Riccioli (Bologna, 1651). Having discarded the Earth-centered Universe of Ptolemy, but not yet ready to embrace the Sun-centered Copernican model, Riccioli depicts the scales as tipped in favor of a compromise model proposed by Tycho Brahe in which the Sun and Moon orbit the Earth, while the other planets circle about the Sun.

major players in every cosmic interaction. He proposed new equations to describe the curves and bumps that characterize the geometry of space and the trajectories of both mass and light. These equations also rewrote the roles of space and time—allowing for a Universe that changes over time. Even Einstein did not want to accept this most incredible prediction of his new theory.

Time has always played a critical role in the machinations of the heavens, but it was believed to be a cyclical role, one that delineated the seasons and predicted the positions of the planets. Yet the equations that formed the architecture of Einstein's new model were stubbornly insistent, presenting us with a universe that is evolving. Space is either expanding or contracting in a wrenching departure from all that was known or observed or experienced. The Universe has a past and it is headed toward a future, both of which look very different from the present.

This possibility of a time-dependent universe in his theory prompted Einstein to add a *cosmological constant* to his equations—a cosmic fudge factor that would balance the books and hold the Universe in a state of suspended animation, neither expanding nor contracting. It was a mathematical trick motivated solely by the belief that the Universe is static. With the addition of this constant, Einstein missed the opportunity to predict the expansion of the Universe—one of the most radical consequences of his new understanding of space and time. To add insult to injury, this solution was later shown to be unstable.[2] Adding a cosmological constant could produce a static Universe—but any tiny disturbance in the Universe, the smallest change in its size, and the Universe would be off on a runaway expansion or collapse. It's like trying to balance a pencil on its point—it's possible to do, but the pencil doesn't stay upright for long.

Fortunately, general relativity made several key predictions that are independent of the cosmological constant. The first success of Einstein's new theory came with his calculation of Mercury's orbit. Predictions based on Newtonian physics had repeatedly fallen short, unable to explain detailed observations of the path traced out by Mercury, the planet closest to the Sun. With each trip around the Sun, the point of closest approach (perihelion) precesses a small amount—the orbit gets a little ahead of itself.[3] Einstein's calculation of Mercury's orbit using general relativity was in perfect agreement with the observations. The moment when Einstein compared his results with the data has been described as "by far the strongest emotional experience in his scientific life, perhaps in all his life."[4] He knew he was onto something big with his new theory of gravity.

What ultimately rocketed general relativity to the forefront of both the scientific and the public spheres was of a completely different nature. A new theory must successfully explain the existing data (such as the orbit of Mercury), but in addition it must go beyond known phenomena to make new predictions. General relativity's elevation to a grand and revolutionary model of our Universe occurred in 1919, when the successful eclipse expedition led by Sir Arthur Eddington confirmed one of the theory's major predictions—the bending of light by a massive object—and forever changed the way we view the cosmos.

WARPED SPACETIME

General relativity is a rich and complex theory, but its beauty lies in the simplicity of its basic premise: mass and energy warp spacetime; the resulting curved spacetime dictates how mass and energy move.

Mass and energy are mentioned together because they are actually different incarnations of the same beast; the energy of an object at rest (a marble, for example) is its mass; the energy of a moving object is a combination of its mass and the energy of its motion. Light, which has no mass, is pure energy. And all of these forms of energy can be transformed into one another.

The key ideas of general relativity are well understood and have been repeatedly verified by experiments and observations:

- Mass and energy warp spacetime.
- Objects move around in the warped spacetime along the "shortest path."
- The shortest path in a curved space is no longer a straight line.
- Light always moves along this shortest path, and thus light travels through space along curved paths.

Further, general relativity also makes some fantastic statements about the entire Universe. Because space and time respond to the presence of mass and energy, the overall Universe—which means all of spacetime—reacts to all of the mass and energy that are contained within it. Simply put, general relativity implies that

- In general, the Universe will be either expanding or contracting.
- The evolution of the Universe—the history of its expansion or contraction over time—will depend on its composition.

The preceding is a very short summary of a very difficult subject and should leave you feeling a little bit dizzy. It may help to follow the logical steps that led Einstein to make these claims. He began with trying to understand light and how it propagates through empty space. The result of this exercise—which effectively removed the word *simultaneous* from the dictionary of physics, for reasons that should be clear by the end of this chapter—is now known as the theory of special relativity. The moniker *special* derives from the fact that this limited version of relativity theory completely omits any mention of gravity, which Einstein realized was going to complicate matters. (Physicists are experts at leaving out the hard parts at first, then adding them in later.) The successful extension of the theory of relativity to include gravity is known as *general relativity*, and without this full theory we would not be able to make sense of the Universe in which we live.

SPECIAL RELATIVITY

Einstein had a banner year in 1905. With the publication of six relatively short papers[5] (as compared to the lengthy tomes often published by physicists today), he launched two new branches of physics—quantum mechanics and relativity—and took over the helm of a third: the field of statistical mechanics. Quantum mechanics revolutionized the world of the very small; Einstein's explanation of the random motions seen in liquids (Brownian motion) using statistical mechanics was instrumental in confirming the atomic theory of matter; and his theory of special relativity reinvented the role of time in all of our calculations.

It had been assumed that time is like a metronome with a single setting; quietly ticking at even and unchangeable intervals, calling out a strict cadence that marches us all inexorably forward. Einstein said no: the measurement of time depends on who holds the clock.

According to special relativity, the life span of a person—or a particle—moving at nearly the speed of light will appear longer to a person watching from a stationary viewpoint than to someone traveling alongside the moving person. This may sound like science fiction, but we can and have observed this effect. Start with a particle that has a well-measured lifetime. (Heavy particles in general do not exist forever, but decay into their lighter counterparts after a certain amount of time. Muons spontaneously decay to an electron plus two neutrinos. The electron, which has no lighter particle to which it can decay, is stable.) A muon, for example, lives on average for only 2.197 microseconds (millionths of a second). Create a bunch of (stationary) muons in the laboratory where you are standing, and time how long they stick around before decaying. Not surprisingly, you will find an average lifetime of about 2 microseconds. Next, measure the life expectancy of another bunch of muons, identical in every way except that they are zipping past you at an extremely high speed (0.99875 times the speed of light, for example). According to your watch, the average lifetime of these muons is almost 44 microseconds—20 times longer than those that aren't moving.

It's all relative, though—hence the name of the theory. Time is not the same for everyone, and two watches that are moving at high speeds *relative* to one another will not agree. The muons that are racing along at close to the speed of light don't think they've lived any longer than their cousins that stayed at home. And if you and your watch were booking along at the same insanely high speed as the moving muons (which is not yet possible—it takes a lot of energy to accelerate a single particle to such speeds; accelerating an entire person, or better yet a rocket ship that could carry the person, is beyond our technology at the moment), you would measure a decay time of 2.197 microseconds.

The measurement of distances in special relativity also depends on who is doing the measuring. Suppose that as you zip along in your muon-chasing rocket ship, an outer-space shoe store located aboard a rocket passes by, traveling in the opposite direction. The clerk measures the length of your foot, held pointed in the direction of your flight, and declares that you are a size 8, when you know you've always been an 8½. Further, the size 8 shoe that the clerk holds up for your inspection (parallel to the direction of your motion so that you can see both the toe and the gorgeously styled heel) looks like a size 7½ to you. (Your rocket and the shoe store must be passing each other at a relative speed of roughly one-third the speed of light. For comparison, the fastest spaceship launched by NASA to date—New Horizons—is headed to Pluto at about 0.000045 times the speed of light, 22,000 times slower.)

Einstein found a way for you to get shoes that fit. One of the key concepts in special relativity is that there is no unique way to define zero speed in the Universe. It's equally as valid for me to claim that I'm standing still and you're moving away from me as it is for you to claim that you're standing still and I'm the one moving away. Regardless of who is moving, there's still only one shoe size that will fit on your foot. (Or, as Einstein would put it, the laws of physics must be the same for both of us.) What the equations of special relativity do is translate between two of us who are moving with respect to one another. Special relativity allows the shoe salesperson to measure your foot as you speed by and calculate your true size on the basis of how fast you were moving. Likewise, special relativity allows you to calculate the lifetime of muons from their point of view, whether they are in your laboratory or speeding away from it.

Einstein developed the equations of special relativity beginning with two important assumptions. First, the speed of light (in empty space) is always the same. It has a constant value, no matter who is measuring it or how fast that person is moving. Second, the laws of physics must be the same for everyone. These laws do not depend on how fast you're moving (as long as you're moving at a constant speed and not speeding up or slowing down).

Imagine, as Einstein first did, that the speed of light does depend on the motion of the person measuring it. This would mean that if you were traveling in the same direction as a beam of light, and going as fast as the speed of light, you should be able to look at the light beam and see it not moving at all (just as objects inside your car—the coffee mug in your cup holder or the map on the front seat next to you—do not appear to be zipping along at 60 miles per hour, since you are traveling together at the same speed). But light is pure energy—it has zero mass—so if it isn't moving it doesn't exist. There is no way to stop light and see what it looks like. The only way out of this conundrum is to insist that light always moves at the same speed (in empty space), no matter who is measuring it.

Einstein's second assumption is easier to understand. Special relativity states that there is no unique speed against which all observers should calibrate their speedometers—all (constant) motion is relative, and thus no observer can claim to be moving at a special speed. If there is no way to distinguish between different moving frames of reference—viewpoints—then the laws of physics must operate in exactly the same way in all.

Armed with his two postulates, Einstein played "what if." What if (1) the speed of light is always the same (since a light beam that isn't moving makes no sense), and (2) the basic laws of nature are the same for everyone? What does this imply about how the world works? Einstein encoded these assumptions into the mathematical model of how light and particles move, and then cranked the figurative handle of this model to see what it produced. What he found was that these two seemingly innocuous postulates completely overturn our concept of time—it is not the absolute we had thought it was. If the speed of light is constant, then time is not.

The easiest way to see this is to head back out into space and watch a few games of laser tag (see Illustration 2.2). A laser pointer emits light, so the beam from any laser pointer travels at the speed of light. The game is played on board a large space station while a referee watches from a second space platform. The space station and platform are moving past each other at high speed. The captain of the blue team and his counterpart on the white team stand back-to-back in the center of the space station. At exactly the same moment, they fire at the last remaining foot soldiers of the opposing team, who stand paralyzed at opposite ends of the ship, exactly the same distance from the captains at the center. Observing that both soldiers drop at the exact same moment, the two captains turn and shake hands, declaring a tie.

"Not so fast," says the referee. He watched the final play but saw the person at the rear of the space station (as it receded from him) drop first. He insists that the blue team won. From his point of view, the targets moved between the time the captains fired and the instant at which the soldiers were hit. The white soldier at the back had moved closer to the point in space where the captains were standing when they flashed on their lasers—the blue soldier at the front had moved farther away. Because the light aimed at the white soldier traveled a shorter distance but at the same speed, the white soldier was hit first and thus the blue team won.

"No way!" shouts the white captain. "The two soldiers didn't move at all, but stayed fixed in the spots on opposing ends of the ship. The laser light traveled exactly the same distance, at the same speed, striking both of them at the same time, and thus the game ended in a tie."

Who is right—the referee or the captains? They all are. The mea-

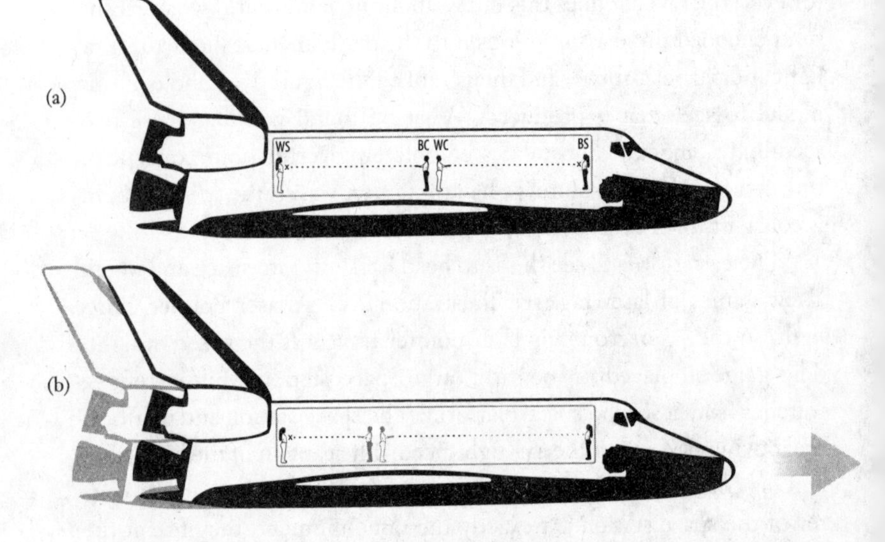

ILLUSTRATION 2.2 Spaceship laser tag. (a) A game of laser tag as seen by the two captains (BC = blue captain; WC = white captain), who are standing back-to-back at the center of the ship. Each captain has fired on the last remaining soldier of the opposite team (BS = blue soldier; WS = white soldier) at the same point in time, with dotted lines indicating the path of the laser light. The captains see the light hit the soldiers at the same time. (b) The same event viewed by a referee on board another spaceship as the game ship zooms by him. The ghosted images of the ship and captains show their positions at the time the lasers are fired; the solid images are the positions of the ship and soldiers when the referee sees the laser light hit the soldier in back. According to the referee, the distance each laser light has traveled is the same (solid dots). During the time it takes the light to travel this distance, however, the ship and the soldiers have moved so that the referee sees the light hit the white soldier before it reaches the blue soldier. (Effects are greatly exaggerated.)

surement of time depends on who is doing the measuring. In other words, the referee's clock and the captain's clock, both of which measured the time between the same two events (in this case the time between the firing of the lasers and the moment when the white soldier collapsed) registered a different time interval. Further, this inability to agree on time intervals also makes it impossible to state that two events are simultaneous. Two events that occur at the same time according to one person will appear to happen at different times according to another person who is moving with respect to the first person. Of course, none of these results are detectable unless the spaceship is moving at close to the speed of light and the clocks are extremely precise—which is why we got along so happily without special relativity for so long. Its effects are in regimes that are far outside most of our everyday experiences, but these effects are very real, with very real and important consequences.

There is something all observers can agree on, however. Einstein broke new ground when he linked a combined measurement of space and time to create a new invariant quantity—a combination that will be the same for everyone. Neither space nor time has the absolute quality we had previously assumed—no two people who are moving with respect to one another will agree on the distance between two points or the time between two events. However, a combination of the distance and the time between two events, known as the *spacetime interval*, is the same for everyone. Time no longer stands out as a distinct and unique quantity, but must be thrown into the pot with its spatial brethren. The underlying unit of measure in the Universe is not an inch or a second, but a unique combination of these.

There are myriad mind-twisting aspects of special relativity, including the famous twin paradox in which we substitute people (twins) for the muons and send one of them racing away from Earth in a yet-to-be invented rocket that can reach velocities of almost 186,000 miles per second. If the adventurous twin zips out to take a closer look at Pluto and then returns home and lands safely on planet Earth, she will find that she is now younger than her more timid sibling. Experiments done with very sensitive clocks placed in airplanes circling the Earth, or with particles created at accelerator laboratories such as Fermilab, have verified these incredible predictions of special relativity to very high precision.

Einstein, however, was just getting warmed up. Next on his list was gravity—how did this fit in to his new theory of relativity?

ADDING GRAVITY

The extension of Einstein's theory of special relativity to accommodate gravity did not proceed in a smooth and direct fashion. The path to this more inclusive theory was long and difficult, and for very good reasons. Including gravity would ultimately demand additional mathematical tools that Einstein did not yet possess. More important, incorporating gravity into his theory of relativity would require the overthrow of an idea that seemed unchallengeable. Special relativity operates on the premise that space is flat, a mere backdrop for the real action in the Universe: *flat* meaning that the shortest distance between two points in space is a straight line and that all the rules of basic high school (Euclidean) geometry are valid; and *backdrop* in the sense that events happen at a certain place—regardless of how we assign coordinates to space (street address, longitude and latitude, or stellar coordinates), it's just a way of saying where the events happen. All of which seems perfectly reasonable.

Wrong. The key insight that propelled Einstein to a new level of genius was that space cannot be relegated to the background—it is a changeable and changing entity that is as dynamic and active as anything that moves within it. This shift in our worldview is akin to the plight of the shipwrecked sailor in the old cartoon who thinks he has landed safely on a small, deserted island—until the island rises up out of the water and the eye of the whale blinks open. The foundations of our world have revealed themselves to be a very different beast than we assumed and, like the sailor, we're about to be taken for a ride.

THE PRINCIPLE OF EQUIVALENCE

The final version of general relativity was completed in November of 1915, 8 years after the 1907 paper in which Einstein had published his first attempts. In the intervening years he had published several other papers on this topic, some of which he himself later showed to be incorrect. In fact, his own frustration with the process is summed up in a letter he wrote in December of 1915, shortly after finishing his new theory: writing about himself in the third person, Einstein notes, "That fellow Einstein suits his convenience. Every year he retracts what he wrote the year before."[6]

Although we won't follow Einstein's exact route from special to general relativity, highlighting a few of the key breakthroughs can be helpful in understanding the final results. The first critical insight came almost immediately. In 1907, as he was thinking about how to modify Newton's laws of gravity, Einstein performed one of his thought (*Gedanken*) experiments. In free fall (off the roof of a house, in Einstein's example), what kind of gravity is experienced? The answer, he decided, is none. Someone, or something, in free fall is in a state (a frame of reference) with zero gravitational field.

That sounds, to be honest, a bit nutty. Obviously, we think, the person is falling because of gravity, so how can Einstein claim that the person is experiencing no gravity? We need to stick with him, and his falling guinea pig, for a bit longer. Imagine that the person in free fall has no external cues—she's inside a big box, for example, and can see only what is inside the box, and thus also in free fall with her. (We'll ignore the unpleasant end stage of the experiment, when she and the box hit the ground, and assume that the fall lasts sufficiently long that she has time to look around and conduct a few mini-experiments).

Asked what gravity is, an 8-year-old will often pick up the nearest block or stuffed rabbit or raisin and let it drop, proudly proclaiming that it falls because of gravity. But what happens if the person in the freely falling box releases the cell phone she's holding? Nothing. As viewed by the person in the box, it doesn't fall but simply floats beside her. She experiences no gravity. Anything and everything in the box (including the box itself) is falling freely, and thus to her appears to float in midair as if there were no gravity.

To those of us watching the box as it passes outside our window on the thirty-fifth floor, the box, and all of the items inside of it, are falling at the same rate (ignoring any air friction). Gravity affects all masses in exactly the same way—everything falls at the same rate of acceleration. And in the same direction. Unlike an electric field, which we can probe by tossing in an object with a negative or positive charge and watching which way it moves, nothing has a "negative gravitational charge." Nothing falls up, away from the center of the Earth, because gravity is always attractive. Ignoring air resistance, a bowling ball, a marble, and a peppercorn dropped simultaneously from the top of the Eiffel Tower will all hit the Parisian sidewalk at the same time. Standing on the sidewalk, we see them all speeding toward us, but to an ant riding along on top of the bowling ball, the marble and the peppercorn are simply suspended in air next to it.

Einstein did not have the advantage we have today of seeing this effect in action. Astronauts orbiting the Earth in the space shuttle or the space station are, in fact, in free fall. Tugged by the Earth's gravity, they are falling freely toward the Earth, even as they are also moving parallel to the surface of the Earth. The resulting path is a combination of the forward motion that was given to the shuttle by its thruster engines in order to place it in orbit in the first place, plus the downward (toward the center of the Earth) acceleration due to gravity. This curved path keeps the shuttle at a fixed distance above the Earth's surface, but it is still free fall. The images of toothpaste or water or space suit gloves that appear to float in midair, beamed back to our television screens from aboard the shuttle, are real-life visualizations of Einstein's thought experiment.

For Einstein, the realization that a person in free fall has no way of knowing if she is accelerating due to gravity (falling freely) or in a gravity-free zone meant that the two conditions are equivalent. If there is no way to tell whether you are experiencing zero gravity or are in free fall, then the two states must be physically equivalent and the laws of physics must be the same in both. Einstein's key insight in 1907 was understanding that gravity is inextricably linked to understanding how to connect physics between reference frames that are accelerating or decelerating—speeding up or slowing down with respect to one

ILLUSTRATION 2.3 European Space Agency astronaut Pedro Duque on board the International Space Station in 2003. Three balls of different materials appear to float in space in front of him. All three balls, along with Duque, the space station, and everything on board, are freely falling toward the Earth—a situation that is equivalent to zero gravity.

another. His first formulation of general relativity was based on the assumption that gravity and (uniform) acceleration are equivalent—an assumption known as the *equivalence principle*.

The equivalence principle offered the window into understanding gravity for which Einstein had been searching. Gravity might be a mystery, but acceleration was something he knew how to deal with.

Since this is still relativity theory, the next question is, accelerating with respect to what? In other words, an apple falling from a tree appears to an observer standing beneath the tree to be accelerating toward the ground—but is the apple accelerating or is the observer? Special relativity states that there is no special velocity in the Universe—that is, there is no way to define a particular person or object that has zero motion, against which everyone else's motion must be measured. However, it is possible to pick out a special acceleration. A person (or apple or rocket ship or planet) that is freely falling due to gravity is defined to have zero acceleration. They are experiencing no force—so no acceleration. Zero acceleration means that they are neither speeding up nor slowing down, so they must be moving at a constant velocity, just as in our discussion of special relativity. Therefore, all the laws of physics must still hold.

Let's return to our unfortunate friend in the box. She has no view of the outside world, and thus there are no external clues to tip her off to the fact that she's falling. She decides to try a little experiment and pulls her laser pointer out of her pocket and aims it at the other side of the box. When she flashes it on, she sees the red laser dot on the box wall directly across from her, just as she expects (see Illustration 2.4a). It's also what we would expect her to see on the basis of the equivalence principle—the physics inside the box must be the same as in any other environment that is in constant motion. (If the light didn't follow a straight

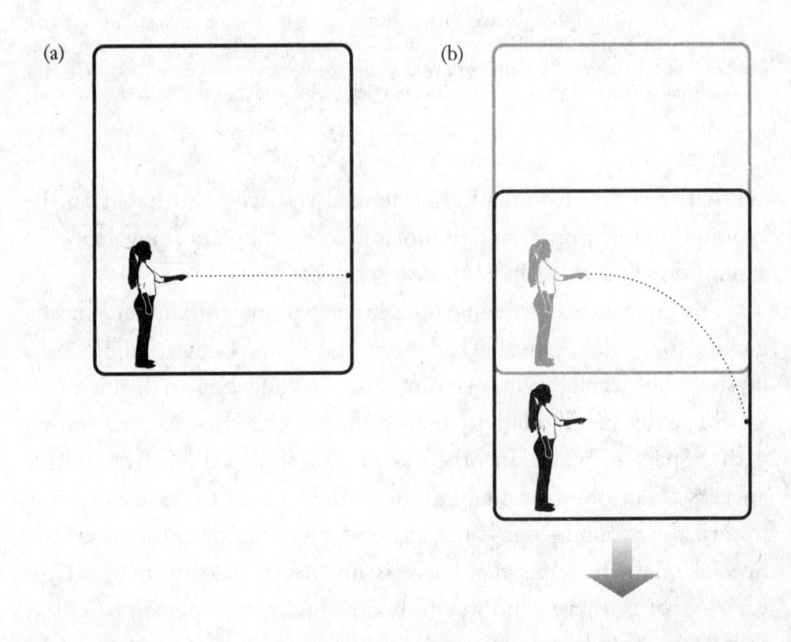

ILLUSTRATION 2.4 Light in free fall. (a) Enclosed in a falling box with no windows (and thus no external clues as to whether she is freely falling in a gravitational field, or in a zero-gravity environment), our observer fires a laser pointer at the opposite wall and sees the light hit the opposite wall exactly where she expects. (b) Seen from outside the box as it falls past us (assuming that we can see through the walls), the light in a gravitational field follows a curved path. The position of the box and the person inside are shown at the time the laser pointer is turned on (ghosted view) and when we see the light hit the opposite wall (solid).

path across the box, our falling experimenter would be able to tell that she was falling, in contradiction with the equivalence principle.)

What, however, does this experiment look like to those of us outside the box, watching as it falls past us? The path of the laser beam from our point of view is not a straight line—it instead follows a curved path as shown in Illustration 2.4b. Because the box is falling, by the time the light reaches the wall, the wall itself has moved. In order to hit the box wall (which is freely falling) at just the "right" point, according to the observer inside the box, the laser light—which has no mass—must also be in free fall. Light responds to the presence of gravity in exactly the same way that the person in the box (and the box itself) does.

When Einstein conducted this part of his Gedanken experiment, he found that his result had major implications for the behavior of light. Light energy must also be subject to gravity, in exactly the same way that any massive object is—in the presence of a gravitational field, light follows a curved path.

The key steps in Einstein's argument can be summed up as follows:

- Gravity is equivalent to acceleration.
- Free fall can be defined as zero acceleration.
- Thus, a freely falling observer is equivalent to an observer in constant motion (and no gravitational field), so they must not be able to tell that they are falling.
- Therefore, gravity must affect light just as it does mass (or the observer could find some clever way of using light to tell that she was falling); that is, light follows a curved path in the presence of gravity.

Gravity's effect on light is not just theoretical—we have seen it in action. Because of gravity, light will lose energy as it moves away from a massive object. Just as a ball loses energy as it is thrown up in the air (working against the gravitational pull of the Earth) so, too, will a beam of light lose energy as it travels from the surface of the Earth to a higher altitude. This energy loss is known as *gravitational redshifting*.[7] For light, a change in energy translates into a change in color. What our eyes detect as blue light is light with a higher energy than what we see as red light. Point a beam of blue light straight up away from the Earth, toward someone hovering overhead in a helicopter, armed with a sensitive light detector. The apparent color of the beam will be different when it is detected at high altitude, shifted ever so slightly toward the red end of the spectrum—that is, "redshifted."

Even stranger are the implications of the equivalence principle for clocks. Gravity alters the rate at which clocks run—the stronger the gravitational field, the slower a clock ticks. To see why, place a set of clocks on board a rocket ship—one in the front of the ship and one in the back. Start with the rocket far out in space (so that there is no gravity around) and the rocket at rest. The two clocks are not moving relative to one another—they're both on the same ship—so the clocks will agree, ticking away at the same pace. If the clock in front ticks every nanosecond and sends a pulse of light to the clock in the back with each tick, the back clock will receive a light pulse every nanosecond. It takes a certain amount of time for the light to reach the back clock, but the interval between ticks detected by the back clock will still be one nanosecond.

Now turn on the booster rockets so that the ship begins to accelerate. The front clock continues to emit a light pulse every nanosecond, but the speed of the ship is continually increasing so that by the time the light is detected, the ship is moving faster than it was when the light pulse was sent. The back of the ship accelerates toward the arriving light—and the pulses arrive closer together. This increase in the frequency of the arriving pulses is known as the Doppler effect. The basic idea can be demonstrated with an automated ball machine that spits out a tennis ball every 2 seconds. If you stand still at the baseline, the ball will bounce off of your racquet at the same rate—once every 2 seconds. But run toward the machine and you'll find that a ball hits your racquet with a higher frequency—you'll have less time (less than 2 seconds) between shots. Exactly how much less depends on how fast you can run. The case of the accelerated rocket is only slightly more complicated. Unlike you and the ball machine, the two clocks never get any closer together. But the change in the speed of the ship between the time the light pulse is emitted and the time it is detected results in a *net* velocity of the back clock with respect to the front clock, so that the detected light pulses arrive at a faster rate than they are emitted, just as you get hit more frequently with the tennis balls.

To someone stationed at the back of the ship, the front clock appears to be running fast. The opposite is true for the back clock viewed from the front of the ship. The front is accelerating away from the light emitted by the back clock, so the light pulses from the back clock take longer to arrive at the front and an observer at the front will claim that the back clock is running slow. Note that both the person in front and the one in back agree: the front clock is running faster than the back clock, even though the clocks are still not moving with respect to one another.

Now apply the equivalence principle, which states that those on board the accelerating ship have no way to tell if they are accelerating or experiencing the effects of gravity. Thus, if the same rocket ship is sitting on the launchpad on a planet whose gravitational pull is equivalent to the acceleration of the moving ship, nose pointed skyward, the astronauts inside the ship will see the same difference between the two clocks. The clock at the top (front) of the ship will be running slightly faster than the one at the bottom. The stronger the tug of gravity (gravity is stronger on the surface of the planet, weaker at higher altitudes), the slower the clock runs—gravity causes clocks to tick more slowly.

This effect of gravity on clocks has also been observed, and it has important real-world consequences, especially for anyone climbing on board an airplane. White-knuckle fliers might distract themselves during their next landing in foggy weather by estimating the difference between a clock on the Earth's surface and one 12,000 miles above it, where the gravitational potential is about four times weaker. Airplane pilots rely on global positioning systems (GPSs) to accurately determine their position in bad seeing conditions. GPSs calculate precise locations by comparing signals from clocks perched on several different satellites—satellites that are in orbit more than 12,000 miles above the Earth's surface. Because the strength of the Earth's gravitational field decreases with distance from the center of the Earth, clocks at this altitude run faster (where gravity is weaker) than clocks on the surface of the Earth. There is also a smaller correction because the satellites are moving at a speed of about 8,700 miles per hour relative to the surface of the Earth. According to special relativity, this (relative) motion will decrease the rate at which the satellite clocks tick. Overall, the combined corrections add up to about 38 microseconds (faster) over the course of a day—1,000 times larger than the precision needed to determine the position of an object to within 15 to 30 feet or so. Failing to include special and general relativistic effects in GPS calculations of your position could make for an unpleasant landing.

BEYOND TIME

Time is only half of the story of general relativity, however. In fact, when Einstein reached this point in his development of the theory, he calculated how much the mass of the Sun would bend the light of a distant star—and got it wrong. His answer was too small by a factor of 2.[8] The problem was that, up to that point, he had considered only the effects of gravity on time: by altering the rate at which clocks run, gravity changes, or warps, the passage of time. Einstein had yet to consider the possibility of a similar warping of space.

Two additional insights were necessary before the full theory of relativity could be realized. First, Einstein recognized that he needed to accommodate variations in the strength of gravity. Without stating so explicitly, all of the thought experiments we've discussed until now have assumed that gravity (or acceleration) is constant and the same everywhere. This is, of course, not true in general. As we travel through the Universe, the gravity we encounter changes as we move closer to or farther from objects with more or less mass. The equivalence principle, which was crucial to the development of general relativity up to this point, does not hold for regions of space large enough to encompass more complicated gravitational situations.

Suppose we fashion a large rectangular space station (see Illustration 2.5), with extremely strong and rigid walls, and a length that is an appreciable fraction of the Earth's circumference. Position the station 50,000 miles above the surface of the Earth so that its length is parallel

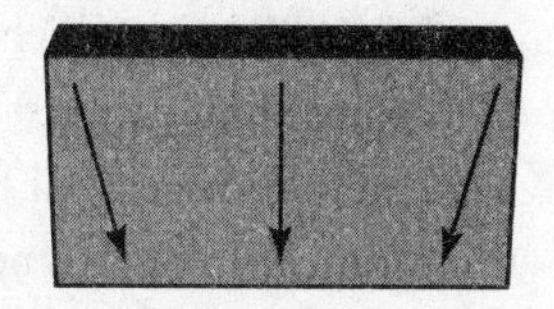

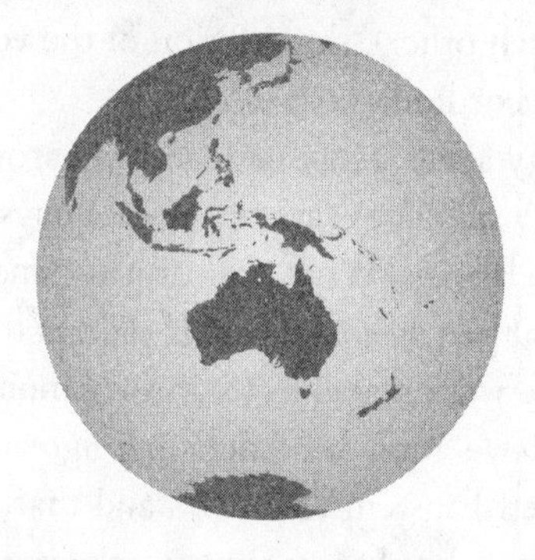

ILLUSTRATION 2.5 Tidal Forces. An extremely large space station in free fall toward the Earth. The arrows indicate the direction of the force of gravity at three locations within the space station—because the station is so large, and the tug of Earth's gravity is toward the *center* of the Earth, these directions are not parallel. Objects (and people) inside of the space station will fall along the directions indicated by the arrows, and thus their positions relative to one another (and the sides of the space station, assuming that it remains completely rigid) will change over time.

to the Earth's surface, offer two billionaires the opportunity to conduct a space experiment, and place them at opposite ends of the station. Now let go. As the space station falls freely toward the center of the Earth, the occupants are allowed to compare notes, and a second trip into space is offered to the first one who figures out which way they are falling. While one of the volunteer space cadets spends his time pushing off walls and turning somersaults, the other very cleverly releases a marble, which floats in midair beside him, and then stays very still. He notices that, as time passes, he and the marble are drifting toward the center of the station and, ever so slightly, toward one of the four lengthwise walls. He immediately points to the opposite side, declares that the Earth must be in that direction, and claims his prize.

In this experiment, the space station is simply too large for the equivalence principle to be valid. The station and its contents are falling toward the center of the Earth, which is a slightly different direction for the billionaire at one end of the station than for his companion at the opposite end. As they each fall toward the center they follow the paths shown in Illustration 2.5, and if the station is long enough and rigid enough (so that it doesn't bend as the ends of the station fall toward each other), the position of the volunteers within the station will shift accordingly.

Only for a very small patch of spacetime around anyone or anything (a sufficiently small box) are the laws of physics inside that patch those of special relativity. What constitutes "small" depends on the gravity in that local part of spacetime. If each of the billionaires in the preceding example were sealed in their own, much smaller space station, they would have a much harder time figuring out what was up. This realization led Einstein to understand that he needed to break spacetime up into tiny bits before applying the equivalence principle—each one small enough that the equivalence principle is valid.

Einstein's second flash of brilliance was his recognition that the geometry of space had to be much more complicated than flat, or Euclidean, space. In 1912, Einstein recognized that the equations he had been staring at resembled those of curved surfaces. With the help of his friend Marcel Grossman, he found the mathematical tools that he needed to develop his theory, switching from the Euclidean geometry of flat space to the much more complicated Riemannian geometry of arbitrarily curved surfaces.

In essence, he made a leap similar to that of transforming still photography into video. A video or movie is created from a series of still images, which are shown in quick succession. In a high-quality video the images are taken in very quick succession, so that each image is just the tiniest bit different from the one before it and the one after it, smoothly transitioning between frames. Einstein's process was very similar. He realized that as an object moves through the Universe (in the most general way—speeding up, slowing down, rotating in circles), its trip can be broken up into many small segments by a series of snapshots of the object in quick succession, one right after the other. Each snapshot in this series represents a little patch of space and time in which the object is moving relative to us with a certain speed. But that speed is different in each photo. Thus, each frame—each little patch—looks like flat space where we should be able to use our equations from special relativity; but each patch requires us to use a different velocity in these equations. Lots of little flat-space patches, but each tilted slightly with respect to its neighbors. Exactly what you would expect if you were trying to cover a curved surface with small tiles—each one flat, but each one angled with respect to its neighbors so as to smoothly cover the curves of the larger surface.

Putting it all together, an object moving through a Universe filled with lumps of matter and energy of all sizes can be viewed as moving through tiny (almost infinitely small), sequential patches of flat spacetime, each of which is slightly different from the one before and the one after. Near a large mass, where gravity is stronger, spacetime is more sharply curved and we need to break it down into many tiny patches in order to smoothly trace out the shape of spacetime in this region. In empty sections of the Universe, spacetime will be essentially flat, and a few larger patches will do nicely. Once we put them all together, spacetime will look like our tiled surface—not flat, but wondrously curved and warped.

ACTIVATION OF THE GRID

Einstein had more work to do—in particular, formulating a dynamic version of the theory, in a sense bringing his curved spacetime to life. The distribution of mass in the Universe is always changing as objects move through the Universe or clump together in new ways. Stars and planets are in constant motion; galaxies grow in size and mass as they cannibalize their smaller neighbors. Einstein proceeded—with some help from friends, especially Marcel Grossman—to construct a mathematical model of spacetime that allowed it to actively respond to this unruly state of affairs. The final result allowed him to explain the recalcitrant orbit of one of the planets, make a new prediction about the bending of light, and offer the world a completely new description of the Universe.

CHAPTER 3

A Cosmic Expansion

General relativity is viewed by physicists as one of the most beautiful theories ever written down, and it can be elegantly summarized in one equation. *Einstein's equation*, as it is known, is shown in Illustration 3.1, simply because it was impossible not to include it. (This equation is actually a compact way of writing down 10 equations, and the small subscripts—Greek letters μ and ν—are stand-ins for directions in space and time.) In these few dots of ink the operating instructions for the Universe are elegantly laid before us.

$$R_{\mu\nu} - \tfrac{1}{2} R g_{\mu\nu} = 8\pi G T_{\mu\nu}$$

ILLUSTRATION 3.1 Einstein's equation. The left-hand side describes the curvature of space; the right-hand side specifies the mass and energy content.

Beauty aside, Einstein's equation basically relates a detailed description of spacetime—how and where it is curved—to the distribution of the matter and energy within it. The left-hand side of the equation describes spacetime, just as a topographic map provides information on how the landscape is sculptured. If you know how to read such a map, you can locate hills and valleys, determine the steepness of the slopes, and plot your hike accordingly. The right-hand side details the location of all of the matter and energy that exist.

Translating from math to English, Einstein's equation becomes

Spacetime curvature ⟷ Distribution of matter and energy

Simply put, it states that space is curved by matter and energy; and that matter and energy move according to how space is warped.

The visual analogy usually invoked to illustrate this equation is to picture spacetime as an infinitely huge rubber sheet. Several limitations of this approach must be kept in mind. First, spacetime has at least 4 dimensions—3 space, plus 1 time (in theories of the subatomic realm known as string or superstring theories, spacetime has 10 or 11 dimensions, but for now we restrict our discussion to the 4 dimensions of everyday experience), but a rubber sheet has just 2 dimensions (ignoring the thickness). Second, the rubber sheet, as viewed from the outside, is embedded in a higher three-dimensional space, and there is the possibility of moving off of the sheet, above or below it. Such movement is not possible when you're talking about the whole Universe—spacetime is all there is and you can't move off or out of it. Finally, in the rubber sheet example, gravity acts on objects on the sheet to stretch the sheet downward (toward the center of the Earth), whereas the deformations of spacetime do not have a direction in that sense. Nevertheless, the rubber sheet can be an extremely useful analogy, providing our limited human brains, which are designed to see and comprehend only three dimensions of space (with an entirely separate concept of time), with a picture of some of the features of a more complex realm.

Imagine that you are a small, flat bug whose entire life is lived out on such a rubber sheet. You cannot fly or jump up off of the sheet, and from your point of view the sheet is infinite in size. Neither you nor any of your ancestors nor any of your descendants ever has or ever will ever reach an edge of the sheet (or even be aware that such an edge exists), no matter how fast any of you move in any direction. And you can't look up or down, away from the sheet. The sheet is your entire Universe. If there is no matter—no stuff—in your Universe (ignoring any mass in your own bug body or those of your buggy friends and relations), the sheet will be smooth and flat everywhere. The shortest distance between any two points will be a straight line, and all of your junior-high geometry will be valid: the sum of the angles of any triangle will be 180 degrees and parallel lines will never meet. If you and your bug friend race from point A to point B on side-by-side parallel paths, booking along at exactly the same speed, your race will end in a tie.

Now add a few objects of interest. Imagine placing a 6-pound bowling ball, a 20-pound bowling ball, and a baseball at separate locations (far from each other) on the sheet. Each of these balls will cause the sheet to stretch, creating a (downward) dimple and curving the surface of the sheet. The amount of the curvature will depend on the mass of the ball (how heavy it is). Your buggy trips across the sheet will never be the same. If you're traveling across the sheet far from any of the balls, you won't notice any difference, but your attempts to find the shortest path between two points in the vicinity of one of the balls will cause you to question your ninth-grade math teacher. Parallel lines

ILLUSTRATION 3.2 A flat, two-dimensional bug world. (a) No mass present; (b) Two different lumps of mass induce warps in the space around them.

will sometimes intersect, and the angles in a triangle will not always add up to 180 degrees.

The very definition of a "straight line" or, more precisely, the shortest distance between two points, will need to be completely rewritten. In flat space, a nice long yardstick could be used to trace out the shortest distance. In the new warped bug world, you'll need a new kind of yardstick—one that can bend and flex, molding and morphing itself to fit the curved contours of the sheet. The technical name for this

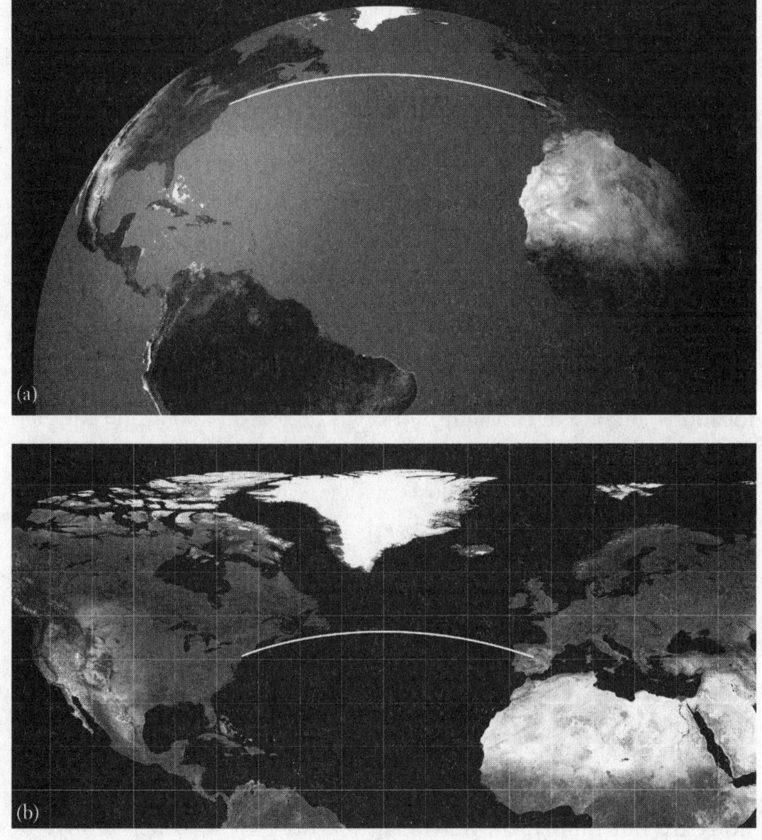

ILLUSTRATION 3.3 The shortest flight path between New York City and Madrid. (a) Drawn on the globe. (b) Shown on a flat projection map with lines of latitude and longitude.

new measuring device is *geodesic*, and anyone who has flown from New York to Spain has probably benefited from our understanding of geodesics. The surface of the Earth is curved (into an approximately spherical shape), and the shortest distances between two points on the surface of a sphere are segments of great circles—the largest possible circles that can be drawn around the Earth. Intercontinental flights are often designed to follow great circles, and thus minimize the flight distance. For example, New York City and Madrid are at about the same latitude, but the shortest distance between them is not a direct east–west route across the Atlantic Ocean, but a path that appears to be curved on our flat maps, arching northward and then back south again.

In general, your nice flat bug world has been transformed and you'll need to develop a new kind of geometry to help you calculate distances (in particular the shortest path between two points) in this new land of curving trajectories. You'll also need to prepare yourself for an even stranger implication of Einstein's equations.

AN EXPANDING UNIVERSE

Spacetime is not only curved, it moves. The stars and galaxies and dark matter that warp and dimple its fabric do not simply sit quietly at fixed locations, but are in constant motion as they orbit about one another, or collapse and coalesce into ever-larger conglomerations as the Universe evolves. Spacetime adjusts its shape in response, echoing the movements of the matter and energy within it—stretching, jiggling, rebounding, and contracting. Move a chunk of mass from one place to another and the spacetime around it will also shift its curves to conform to this change. Stir things up by swirling a couple of black holes around one another and watch the resulting ripples in spacetime that spread across the Universe just like the ripples on the surface of a lake or pond when a stone is tossed into the water. Finally, think really big. What's happening to spacetime on the largest scales? Einstein's equations are not limited to individual lumps of matter such as stars or black holes, but apply equally well to the entire Universe. We know there is matter and energy in our Universe, and according to general relativity the sum total of this mass and energy will dictate the shape—the overall curvature—of the Universe itself.

In 1917, Einstein set out to determine what his new theory had to say about the Universe. He and others began to look for solutions to the equations of general relativity. These equations do not dictate what matter exists in the Universe, or exactly how and where the spacetime is curved but rather say that *if* there is this particular distribution of matter moving around in this way, *then* this is what spacetime looks like and how it moves in response. Or vice versa. Matter and energy influence spacetime, which then influences matter and energy, which in turn influence spacetime, and so on. Finding a solution to the equations means finding a self-consistent description of the distribution and motion of matter and energy and the shape and motion of spacetime. Further, there are lots of such solutions—the trick is to discover those that correspond to what we see happening in our Universe.

These investigations led to a result that made no sense. Einstein had woven time into his theory, and as payback the theory seemed determined to take full advantage of this new freedom. His equations handed him solutions for a Universe that refused to sit still. As soon as he added matter, his model Universe would collapse. Even Einstein, who had cheerfully wreaked havoc with conventional notions of space and time, had never imagined a Universe that wasn't constant. He had observations on his side—distant stars appeared to be fixed in space, and in 1917 no other astronomical observations contradicted the assumption of an unchanging Universe. Our model of the Universe had expanded as humans had explored and mapped it ever farther from our own celestial backyard, but the idea that the Universe itself was expanding was inconceivable.

So Einstein modified his equations. He added a term, known as the *cosmological constant*, specifically designed to keep the Universe from collapsing. At the time, it was about as justified as adding or subtracting a few dollars in order to balance your checkbook. You simply stick in whatever amount will work so that you can reconcile with the bank statement, motivated solely by the desire to zero out the balance, and assume you'll find the reason for the discrepancy later.

The addition of the cosmological constant, which had no real physical motivation of any kind (and, as mentioned in Chapter 2, wasn't a viable long-term solution in any case), allowed Einstein to model the Universe without any ridiculous stretching of space. At least until some fool astronomer went out and measured this stretching 12 years later.

In 1929, Edwin Hubble published the first of his papers on a strange relationship between galaxies and their velocities. Distant galaxies all appeared to be moving away from us, and the farther away a galaxy was, the faster it seemed to be racing in the opposite direction from Earth. A galaxy twice as far away was moving twice as fast; one three times away was moving three times as fast. Exactly what you would expect to see if the Universe was expanding.

Hubble's observations provided convincing evidence that the Universe is not static—that space itself is indeed expanding—exactly as predicted by models of the Universe developed by Alexander Friedmann and Georges-Henri Lemaître from Einstein's original, unmodified equations of general relativity. Upon hearing of Hubble's results, Einstein immediately chucked out his cosmological constant, which he later called his "biggest blunder,"[1] and accepted this revolutionary new view of an expanding and evolving Universe.

THE COSMIC STRETCH

It's hard to picture this expansion—how can space "expand"? Hubble's observations of the galaxies are also consistent with the trajectories of fragments from an exploding bomb. Wouldn't it be easier just to assume that we are sitting at the very center of the Universe, at the ground-zero site of some ancient explosion? Scientists are loath to make such an assumption, and with good reason. It implies that we occupy a unique position in the Universe, and the more we find out about the Universe, the more average our location appears to be. We are sitting on a planet in orbit about a fairly typical star on the outskirts of a fairly typical galaxy. More important, however, all of the observational evidence we have from looking at the Universe in many different ways, from the distribution of the galaxies to the faint afterglow of the Big Bang, is consistent with what is known as the *cosmological principle*.

The cosmological principle states that, over large regions of space, the Universe is the same in all directions (isotropic) and the matter and energy within it are distributed evenly (homogeneous). The word *large* in this context refers to a chunk of space many billions of light-years on each side—there are certainly lumps of matter (otherwise known as galaxies) sprinkled across the Universe, but one multibillion-light-year section of the Universe will look pretty much like any other multibillion-light-year section. In other words, there is no special place or special direction. Further, we can understand Hubble's observations of the apparent recession of the galaxies without claiming a special location. If space is expanding, anyone anywhere in the Universe will see essentially the same picture that we do. Sitting on a planet in a galaxy 10 million light-years away, a scientist looking out at the Universe will also see that all the distant galaxies are moving away from her, and that those farther away are moving away faster.

How can this possibly make sense? It's impossible to draw the exact image for you because it would require sketching an infinite (or unbounded) three-dimensional space that is not embedded in a higher-dimensional space. That can't be done on a two-dimensional piece of paper or as a three-dimensional model fashioned out of clay or plastic, for example. But as scientists, we can move beyond the limitations of our human inability to visualize higher-dimensional spaces, by constructing models out of mathematics. Just as engineers build test models out of balsa wood or metal, physicists build test models out of math—then put them through their paces to learn more about the models. However, we also make use of analogies we can visualize (such as the rubber sheet bug world), which help us illustrate certain aspects of the more detailed mathematical models. We just keep in mind the limitations of these toy models and make sure to return to the more complicated mathematical structure when we calculate the predictions of any theory.

A common analogy used to illustrate the expansion of the Universe is the surface of a balloon—and it's most helpful if you actually

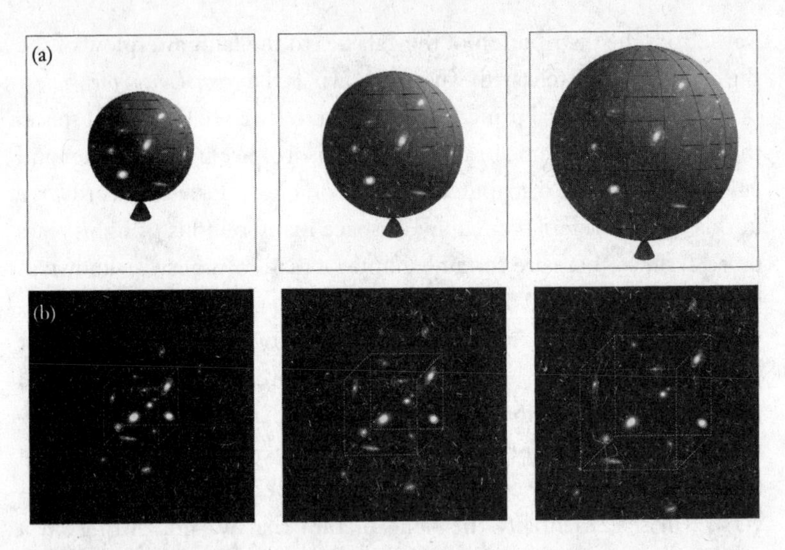

ILLUSTRATION 3.4 The expansion of the Universe. (a) As described in the text, one analogy for an expanding space is the surface of a balloon. Each frame represents a step forward in time; the distance between galaxies drawn on the surface of the balloon increases with each time step as the "space" between them increases. (b) In this three-dimensional example of a cube of expanding space, the cube should be imagined as extending to infinity in all directions. Again, each frame represents a step forward in time, and the distance between any two galaxies increases with time, at a rate that is faster for those that are farther apart.

conduct this mini-experiment. The balloon should be as spherical as possible, and reasonably large so that you can see what's going on. Blow the balloon up until it forms a fairly firm sphere. Put several ink dots on the balloon, evenly spaced about one inch from each other. Pick any two dots and mark them with an extra little dot of color so that you can keep track of them. Inflate the balloon further, then remeasure the distance from one of the colored dots to the other dots. Do the same for the second colored dot. What you will find (assuming the balloon inflates evenly, maintaining a roughly spherical shape) is that it doesn't matter which dot you choose as a starting point, all of the other dots get farther away from it; and the new distance between dots depends on how far apart they were to begin with. For example, if a dot initially 1 inch from a colored dot is now 2 inches away (the distance between them increased by 1 inch), then a third dot that started out 2 inches away from the colored dot is now 4 inches from it (the distance here increased by 2 inches). Assume that it took you 1 second to inflate the balloon. Then the closer dot moved (away from you) at the speed of 1 inch per second, while the farther dot moved away at a speed of 2 inches per second.

This experiment helps illustrate two of the key points in understanding the expansion of the Universe. The surface of the balloon represents space; the dots are galaxies. As you inflate the balloon, the surface area increases just as space increases during the expansion of the Universe. There is no special dot—the view from any of the galaxies (dots) is the same as any other; everything appears to be moving away from you no matter where you are sitting, and the farther it is the faster it appears to recede.

What can be confusing, however, is that the analogy breaks down because the surface of the balloon is expanding into a higher (three-dimensional) space. There is an inside and outside of the balloon. In the case of the Universe, there is no inside or outside. Space—the *surface* of the balloon—is all there is.

Also, the overall shape of the Universe (at least over the part that we can observe) is not curved like the surface of a balloon, but flat, like an infinite rubber sheet—that is, the bug universe, but with the additional twist that the rubber sheet is being stretched evenly in all directions.

A third very important point is that the ink dot on your balloon also expanded, but galaxies do not. Once an object in the Universe gets massive enough, it decouples from the overall expansion. Gravity trumps the expansion, so the Solar System is not getting larger, and neither is the Milky Way Galaxy—the space between the planets is not expanding, nor is the space between the stars in the Galaxy.

It usually takes a while for some of these ideas to sink in, but it's worth struggling with this concept because it provides the foundation for understanding the Big Bang theory and all of cosmology today.

WHO'S DRIVING?

So what does all of this have to do with figuring out the contents of the Universe? Einstein's equations give us the recipe for how space is expanding, spelling out in detail how the expansion is regulated by the matter and energy that exist within it. If we can track the expansion history of the Universe, we can reconstruct the cosmic composition at any given time.

General relativity—gravity—is not a very discerning mistress. It does not distinguish between the different kinds of matter we have discussed (protons, dust, black holes, and dark matter are all considered matter), or between different forms of radiation (microwaves, gamma rays, and ultraviolet light are all the same as far as gravity is concerned). Instead, general relativity lumps the contents of the Universe into three different categories, depending on how they affect the evolution of spacetime.

First there is matter, a category that includes both normal matter and dark matter—any kind of matter with normal gravitational interactions. To be considered matter, it also can't be zipping around at speeds approaching the speed of light—most of its energy must be in its mass, not its motion.

The second category is energy in the form of radiation. Radiation encompasses light of any wavelength, and also very fast moving particles—particles whose energy of motion is much, much larger than the energy associated with their mass. Such particles behave more like radiation energy than like matter.

The final category is dark energy. A cosmological constant would be tossed in this bin, along with any other substance that doesn't dilute away as the Universe expands.

These three basic types—radiation, matter, and dark energy—each have a different impact on the expansion history of the Universe. The Universe began its expansion for reasons we don't yet understand, but Einstein's equations dictate exactly how the expansion will proceed in response to the contents of the cosmos. Both matter and radiation energy cause spacetime to bend and contract and thus help to put the brakes on the cosmic expansion. Radiation, which has a pressure associated with it, is a bit more effective at slowing down the Universe, but the basic mechanism is the same for both. All of the matter and radiation in the Universe is gravitationally attracted to all the other matter and radiation. As the Universe expands, the distance between bits of matter such as galaxies increases. Gravity wants to pull these same bits of matter closer together, in opposition to the expansion. The net result is a slowing down of the expansion of space.

Dark energy is the freaky one. It doesn't slow the expansion of the Universe but gives it an extra boost, causing the expansion to accelerate. In Einstein's equations, the only kind of substance that can generate such a stretching of space is something with a negative pressure. The concept of negative pressure is illustrated in situations right outside your window. In order for water to flow from the root system of a tree to the top branches (against the force of gravity), a negative pressure must exist in the upper reaches of the tree's circulatory system. However, concocting a cosmic substance that has a negative pressure is not easy—matter in the Universe has essentially zero pressure and radiation has a positive pressure associated with it, so whatever dark energy is, it must be radically different from any known substance.

In a Universe filled with only matter, the amount of matter will determine the ultimate fate of the cosmos. If there is less than a critical density of matter, the Universe will go on expanding forever, its rate of expansion slowed by the matter, but never reaching zero. A higher density of matter will result in a more dramatic conclusion: the gravitational self-attraction of the matter will eventually bring the expansion to a halt. Space will stop stretching and instead begin contracting, rushing together at an ever-faster rate until the Universe self-implodes in the antithesis of the Big Bang—the Big Crunch.

The discovery of a dark energy component of the cosmic energy budget put an end to this ability to predict the future course of the Universe on the basis of its content—at least until we discover exactly what dark energy is. If it is truly a cosmological constant, whose energy density stays at the same value forever, the Universe will continue to accelerate, eventually becoming a cold, empty, quiet void, barren of even the simplest structures. If it is not, then all bets are off. The dark energy could decay, diluting its influence until matter once more takes over, or it might grow in strength until it violently shredded every object within the Universe—a cosmic catastrophe that has been dubbed the Big Rip.[2] We simply don't know yet.

In general, the expansion history is dictated by the dominant kind of energy in a particular cosmic era. All three kinds—matter, radiation, and dark energy—appear to have had a shot at running the show. Radiation took the first shift, directing the expansion for over 50,000 years before handing the reins to matter. Matter was in charge for billions of years—a fortunate turn of events from a human point of view—reigning long enough for the Universe to create stars and galaxies and clusters. Then, a few billion years ago, dark energy took control and the expansion of the Universe got its second wind.

During each of these three epochs, the ruling class of energy was in command of both the expansion rate of the Universe and the ability of gravity to gather matter into ever more concentrated structures. Thus, as we backtrack through the history of the Universe, we can search for two signatures: how fast the Universe was expanding, and when and how quickly structures began to form.

The question is how to find these clues. Einstein and his theory of general relativity left us with an incredible picture of the Universe, its undulating curves and warps tracing and tracking the lumps and strands of matter and energy as they move and coalesce over billions of years of cosmic history. Embedded within this theory is also a practical tool for exploring the Universe and ferreting out the nature of its contents. The bending of light that nailed the case for a warped spacetime is more than a scientific novelty. It is also the mechanism behind our most powerful new telescope. And we've now figured out the operating instructions.

CHAPTER 4

Einstein's Telescope

In August of 1914, an intrepid (and ill-fated) group of observers, led by German astronomer Erwin Freundlich, set out to observe a solar eclipse from the vantage point of the Crimean Peninsula in Russia. They hoped to make the first detection of gravitational lensing and thus confirm Einstein's theory of general relativity, which asserted that spacetime would be warped by the mass of the Sun. This dent in spacetime should, according to Einstein, act as an effective lens, distorting our view of distant stars. In particular, a star seen near the edge of the Sun would appear to be shifted out of its usual position in the sky. The distant star itself would not have moved, but its light would travel along a curved path around the warp in space created by the Sun, effectively shifting the observed position of the star on the sky. Because the dazzling light of the Sun overpowers the fainter gleam of stars that lie beyond it, this shift would normally be impossible to see (although Einstein, ever the optimistic theorist, wrote to the director of the Mount Wilson Observatory in California, the largest telescope in operation at the time, to ask if maybe the observers had some neat trick up their sleeves that would allow them to circumvent this little problem;[1] the answer was a rather definitive no). The only opportunity for observing stars so close to the Sun is during an eclipse, when the Sun's light is temporarily blocked by the Moon and the stars once again slip into view.

Unfortunately for Freundlich and his team, they arrived in Russia just in time for the start of World War I. Captured and imprisoned by

Russian soldiers, they were unable to complete their mission—which in retrospect was actually not so unfortunate for Einstein and his fledgling theory. General relativity was still a work in progress in 1914, and Einstein's calculations were off by a critical factor of 2—he had not yet realized that space itself was curved. His prediction for the bending angle induced by the Sun on a light ray passing close to it was too small, and if Freundlich's expedition had succeeded, the observations made during this expedition might have undermined early support for this revolutionary new concept of space and time. Revolutions seldom proceed smoothly from the very beginning, however, and at least in this case there were no casualties—both the explorers and the theory survived.[2] The astronomers were safely back home a month after their capture, and the theory was finalized, with a new prediction, on November 15, 1915.

Over the next few years, war and Mother Nature continued to frustrate attempts to detect the gravitational bending of light during an eclipse, further delaying this critical test of Einstein's new theory. Fortunately, total solar eclipses are not rare—about seven per decade can be seen from somewhere on Earth, if the weather cooperates and the vantage point is accessible. Thus, in the spring of 1919 two teams of British astronomers headed south to set up positions near the equator in order to view the eclipse of May 29, which had the added (and crucial) benefit that the eclipsed Sun would lie in front of a particularly bright star field, a star cluster known as the Hyades.[3] Sir Arthur Eddington ensconced himself and his group on Príncipe Island off the coast of West Africa, while Andrew Crommelin set up camp and camera across the Atlantic Ocean in Sobral, Brazil.

The weather in Africa was wretched, clearing somewhat for only a brief period during the eclipse, and the instrumental setup in Brazil was finicky, as temperature changes affected the equipment and blurred the images, but the combined efforts of the two groups were a success, due in large part to a small (4-inch) backup telescope in Sobral. The observers at Sobral and Príncipe returned home with photographs in which a dozen stars were clearly visible near the disk of the eclipsed Sun. They had obtained the first image ever of the Hyades

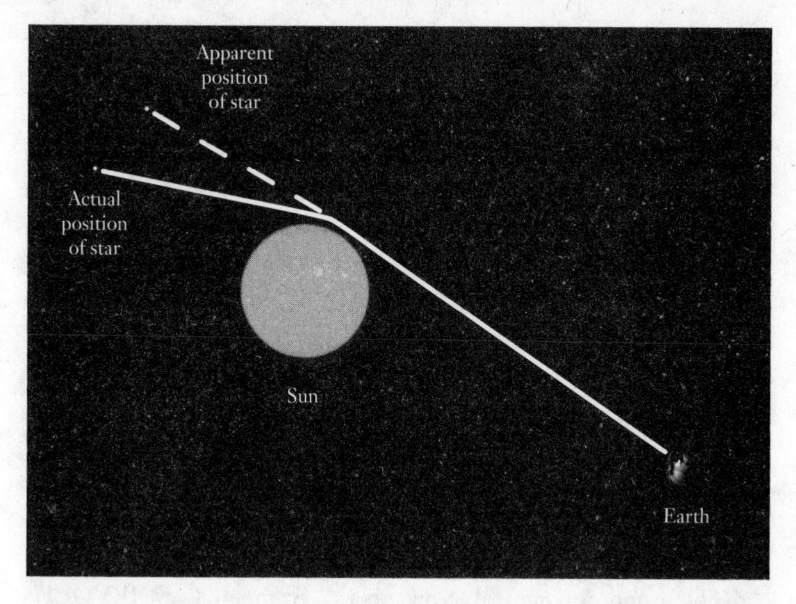

ILLUSTRATION 4.1 Gravitational lensing of a star behind the Sun. The warp in spacetime created by the mass of the Sun deflects light from a distant star and the observed location of the star is shifted.

as seen through the gravitational lens of the Sun's mass. When compared with previous images of the same group of stars taken at night (when the dimple in spacetime created by the Sun would not be between these stars and the Earth, and thus would have no effect on observations of these stars—i.e., *without* the lens), the apparent positions of the stars had indeed shifted. Sir Arthur Eddington's published observations of the amount of the shift agreed with Einstein's calculations of 1.74 arc seconds (slightly less than half of one-thousandth of a degree) to within 20%, and more recent measurements have confirmed this value to an accuracy of better than 1%. The curvature of spacetime had been observed.

As a bit of an aside, Eddington's treatment of the eclipse data has been called into question.[4] He had essentially three sets of observations to work with—two from Sobral and one from Príncipe. The data from the small telescope in Sobral was the highest in quality and implied a shift of 1.98 arc seconds (plus or minus 0.12 arc second). The

Príncipe telescope was blocked by clouds for much of the eclipse, so the data were not as good, and a value of 1.61 (plus or minus 0.30) arc seconds was obtained. The photographic plates from the large Sobral instrument, which experienced "technical difficulties," yielded poor-quality data and found a shift of only 0.93 arc second—much closer to the Newtonian prediction than that of Einstein. Eddington presented all three sets of data to the Royal Society, having decided not to average them because the third set was of inferior quality to the others, for reasons that were well documented. This hardly qualifies as hiding or tampering with data, although a more careful statistical analysis might have been appropriate.

The results were announced on November 6, 1919, at a joint meeting of the Royal Society and the Royal Astronomical Society in London. With amazing speed the news leapt from the scientific cloister to the headlines of major newspapers.

The London *Times* of November 7 trumpeted a "REVOLUTION IN SCIENCE: New Theory of the Universe: Newtonian Ideas Overthrown," and quoted the president of the Royal Society, Sir Joseph John Thomson, as stating, "Our conceptions of the fabric of the universe must be fundamentally altered."[5] Einstein's radical model of space and time had been confirmed, and from this point on his name became synonymous with genius.

A MOST CURIOUS EFFECT

The Sun—or, more precisely, the warp in spacetime that it creates—thus became the first known gravitational lens. The view through this lens was historic—its very existence implied the dramatic overthrow of the known world—but the true power of gravitational lensing as a cosmological tool was not fully realized until long after Eddington's expedition made the headlines. A gravitational lens can create images far more surreal than a simple shift in stellar positions, and can reveal the presence of dark objects that, unlike the Sun, would otherwise remain hidden from view.

The possibility that the warped spacetime of general relativity could lead to strange optical effects was recognized almost immediately—and dismissed as a useless theoretical footnote even more quickly. In 1912, while he was still struggling to complete his new theory of gravity, Einstein sketched out the equations for the lensing properties of a star (see Illustration 4.2 on pages 72–73). He correctly calculated that a star could act as a lens, magnifying the light from a more distant star (or other source of light), or even, in some cases, producing a double image of a single light source. He did not publish these notes, however (and at this time he was still using the incorrect value of the bending angle), and it's not known if he shared these insights with any of his colleagues.

Sir Arthur Eddington was also quick to understand the potentialities of a warped spacetime. Even before he left for Africa, he noted that "the gravitational field round a particle will act like a converging lens."[6] Taking the idea one step further, he discussed the production of a "secondary image" on the opposite side of the lens in his 1920 popular-science book, *Space, Time and Gravitation*.[7] And in 1924, Orest Chwolson mentioned the possibility of observing "fictitious double stars" when a distant star is lensed by a massive giant star, also noting that, under the right conditions, lensing could spread the pinpoint light of a single star into a ring of light.[8]

In 1936, at the urging of Rudi Mandl, a Czech electrical engineer, Einstein redid his earlier calculations and published an article in *Science* entitled "Lens-like Action of a Star by the Deviation of Light in the Gravitational Field." In this oft-quoted publication, Einstein outlines the lensing of one star by another, including the creation of a ring of light (he seems to have been unaware of Chwolson's earlier work) now known as an *Einstein ring*[9] and the production of double images. However, he considered lensing a "most curious effect" of no practical use, stating, "Of course, there is no hope of observing this phenomenon directly." He also sent a note to the editor of *Science* saying, "Let me also thank you for your cooperation with the little publication which Mr. Mandl squeezed out of me. It is of little value, but it makes the poor guy happy."[10]

On this point he was wrong, for two reasons. Einstein underesti-

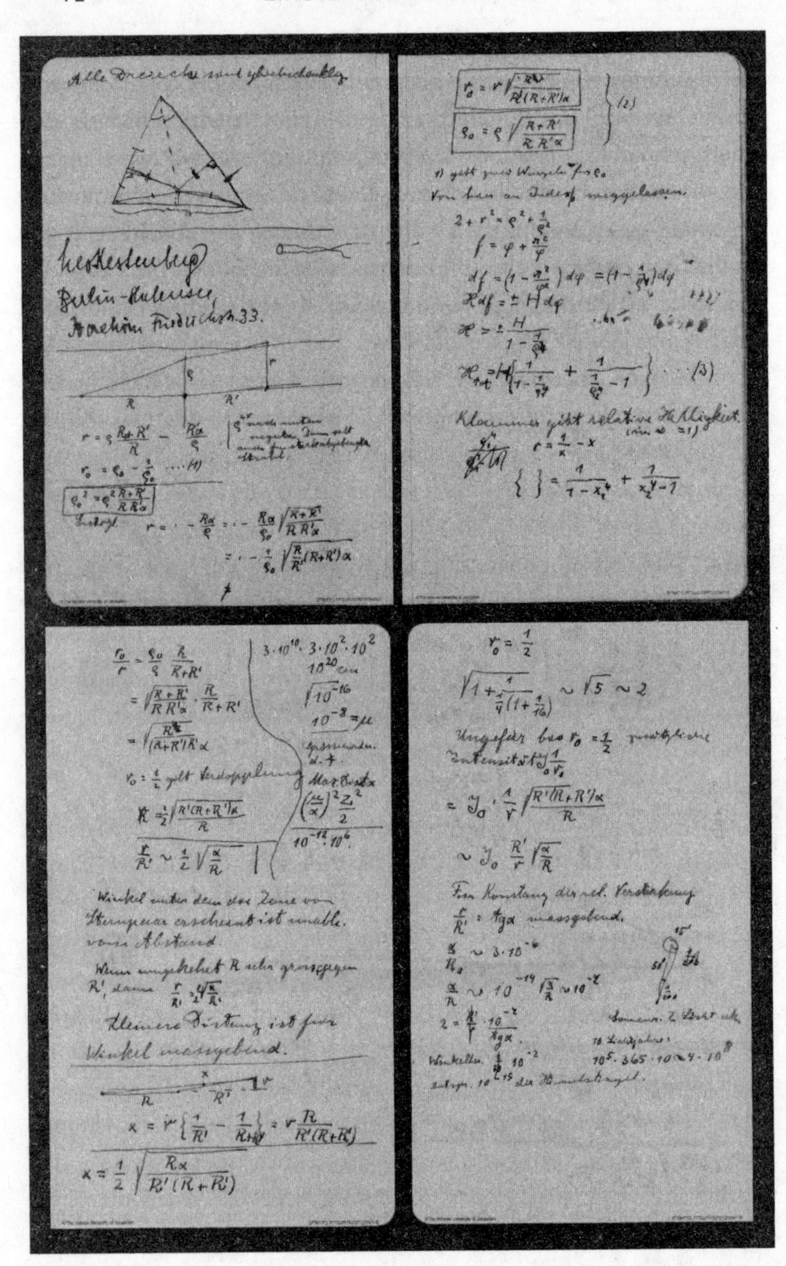

Alle Dreiecke sind gleichschenklig

Berlin-Nikolassee,
Joachim Friedrichstr. 33.

$r = \rho \frac{R+R'}{R} - \frac{R\alpha}{\rho}$

$r_0 = \rho_0 - \frac{1}{\rho_0} \quad \ldots(1)$

$\rho_0^2 = \rho^2 \frac{R+R'}{RR'\alpha}$

$\rho_0 = \rho \sqrt{\frac{R+R'}{RR'\alpha}}$ (2)

Klammer gibt relative Helligkeit.

$r_0 = \frac{1}{2}$

$\sqrt{1+\frac{1}{\frac{1}{4}(1+\frac{1}{16})}} \sim \sqrt{5} \sim 2$

$\frac{r}{R'} \sim \frac{1}{2}\sqrt{\frac{\alpha}{R}}$

Kleinere Distanz ist für Winkel massgebend.

$x = \frac{1}{2}\sqrt{\frac{R\alpha}{R'(R+R')}}$

Für Konstanz der rel. Verstärkung $\frac{r}{R'} : tg\,\alpha$ massgebend.

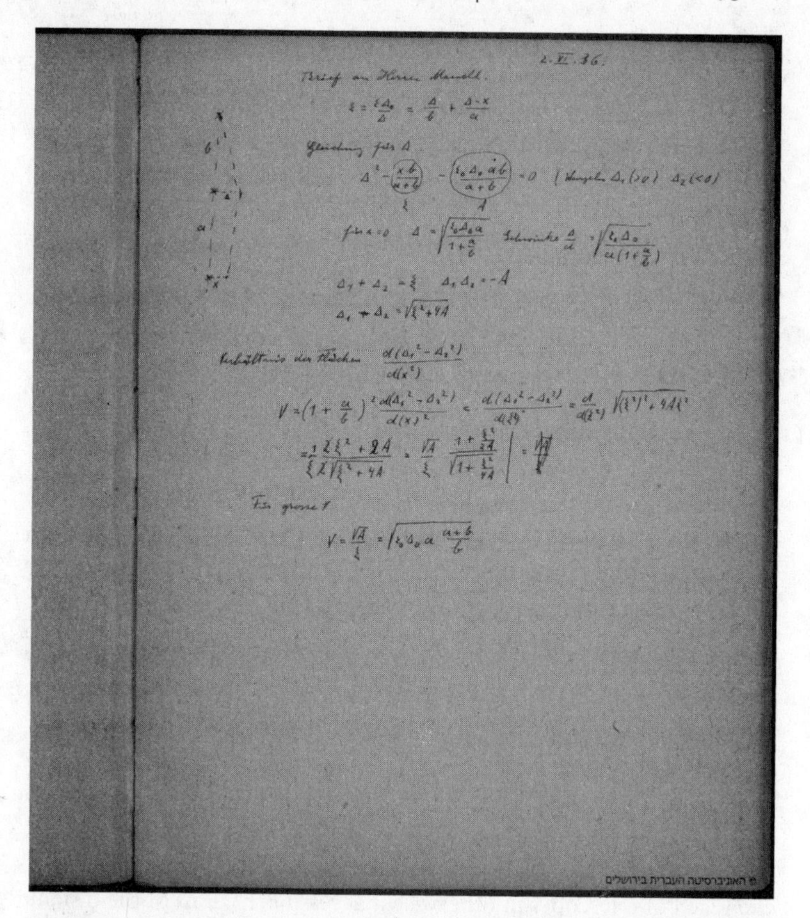

ILLUSTRATION 4.2 Einstein's lensing calculations. (a) Four pages from Einstein's 1912 notebook, written during a visit to Erwin Freundlich in Berlin. (The bottom two pages are the front and back of a single loose sheet found between the two pages in the top row.) Einstein had earlier published a paper on the gravitational bending of light ("On the Influence of Gravitation on the Propagation of Light," *Annalen de Physik* 35, 1911), in which he first suggested testing his theory of relativity by observing the shift in a star's position relative to background stars during an eclipse. In this notebook (starting halfway down the first page), he sketches the geometry of lensing, notes the possibility of a double image, calculates the magnification, and estimates the shift in the apparent position of a star near the edge of the Sun. (b) "Letter to Herr Mandl." Einstein redid his lensing calculations on this page from a 1936 notebook (possibly unaware that he had done a similar calculation 24 years earlier), prior to writing a short letter to the journal *Science* entitled "Lens-like Action of a Star by the Deviation of Light in the Gravitational Field."

mated the advances in technology—larger telescopes, space-based telescopes, and computerized data collection and analysis—that would allow detailed observations of his "most curious effect." More important, he had no way to foresee the current motivation for the expanding use of gravitational lensing. When Sir Arthur Eddington presented his eclipse images in 1919, there were no hints yet of the existence of dark matter and dark energy, and thus no reason to suspect that gravitational lenses would eventually be one of the most powerful tools in a cosmologist's repertoire.

Fritz Zwicky, on the other hand, was right on top of the possibilities. After reading Einstein's 1936 paper (and with a little nudge from a colleague who had also been approached by the persistent Mr. Mandl), Zwicky almost immediately realized that galaxies (which he referred to as "nebulae") would make much better gravitational lenses than stars would. He had earlier provided the first evidence that galaxies were much more massive than had been previously assumed, containing prodigious amounts of dark matter, and he used his estimates of the average mass of a galaxy to calculate the size and brightness of an Einstein ring created by the gravitational lensing of a distant galaxy by another galaxy much closer to us.

In two letters to the journal *Physical Review* in 1937, Zwicky published his calculations and pointed out that gravitational lensing by galaxies would (1) allow further tests of general relativity; (2) extend the reach of telescopes in otherwise inaccessible regions of the distant Universe by magnifying more distant galaxies, and thus "throw very welcome new light on a number of cosmological problems"; and (3) allow a more "direct determination" of the masses of galaxies. He also boldly declared that, if his mass estimates for galaxies were correct, "the probability that nebulae which act as gravitational lenses will be found becomes practically a *certainty*."[11] The italics and emphasis are Zwicky's, who was not in the least shy, humble, or easy to get along with—and in this case he was (once again) right.

It would be a few years before the observers caught up. In the early 1960s, a wave of theoretical papers on gravitational lensing hit the stands (or at least the physics journals) but it was not until 1979, 60 years after the Sobral/Príncipe expeditions, that astronomers found another gravitational lens.

COSMIC TWINS

Quasars are brilliant powerhouses, the most energetic objects known to exist, that can be seen many billions of light-years away. For years after the first quasar was identified in 1963, these strange points of light remained a cosmic mystery, but the evidence now points to a close association with the supermassive black holes that have been found at the center of most galaxies. Supermassive black holes are far more sedentary today than in the distant past, when their younger selves were busy voraciously consuming the matter around them, giving off enormous amounts of light energy in the process. This light energy is intense enough to be seen halfway across the Universe as a quasar.

Viewed through a normal telescope, a quasar looks like a point of light, much like a star. (Hence the name *quasi-stellar object*, which is abbreviated to *quasar*.) However, if a massive galaxy lies directly between the quasar and Earth, what we observe is stunningly different. The mass of the galaxy, which may be more than a trillion times the mass of the Sun, warps the spacetime around it and this warp acts as a lens that spreads the point of light from the quasar into a ring of light that appears to encircle the galaxy—the rare phenomenon known as an Einstein ring.

More often the quasar does not lie exactly behind the galaxy, but just slightly to one side of it. When this occurs, multiple images of the quasar can be produced—two or four identical copies of the same quasar are seen. This unmistakable hallmark of gravitational lensing was a known theoretical possibility in 1979, but the serendipitous discovery of a twinned quasar nevertheless took the observers by surprise.

QSO 0957+561 had originally been noted as a source of radio waves by astronomers surveying the sky looking for hot spots in the radio regime (light with wavelengths much longer than optical). The radio source was not a point on the sky but a tiny region, and within

ILLUSTRATION 4.3 "Twin" Quasars. In the center of this image are two bright objects with (diffraction) spikes radiating outward. These are two images of a single quasar known as QSO 0957+561, which has been lensed by a closer galaxy. The lensing galaxy can be seen behind (just slightly above and to the left of) the lower quasar image. The original discovery of the lensed quasar was made in 1979; this image, taken in 1995 by the Hubble Space Telescope, is a composite of several images in two colors obtained using the WFPC2 camera.

this region were found two optical counterparts—two blue "stellar" objects, labeled 0957+561A and B, about 6 arc seconds (less than two-thousandths of a degree) apart. When these objects were examined more closely, they were found to be the same distance (about 6 billion light-years) away from Earth—a measurement that also definitely marked them as quasars. Stars at such an enormous distance would not be visible.

Finding such a "double quasar" was a rarity, and the two astronomers at the end of the telescope, Dennis Walsh and Bob Carswell, immediately called in a third colleague, Ray Weymann, to take a look at this astronomical oddity. The spectral fingerprint of the two quasars was identical, implying that 0957+561 was not in fact a double quasar, but a double image of a single quasar—a replicating act produced by the lensing action of an intervening (and initially unseen) galaxy.[12]

Zwicky had been right—galaxies do make excellent gravitational lenses. He was also correct in foreseeing that lensing would produce more than pretty pictures.

GRAVITATIONAL LENSING

The deflection of light by massive objects is, in a very real sense, a lens through which we can scan the heavens and search for dark matter and dark energy. Gravitational lensing has the potential to launch a new scientific revolution by making it possible to map out the invisible sector of the Universe. It is a new and powerful telescope—*Einstein's Telescope*.

At the heart of Einstein's Telescope is the lens, which can be as simple as a single massive object such as a star or black hole, or as complicated as a cluster of galaxies—a collection of lenses that can produce beautifully bizarre, almost kaleidoscopic images. But in all cases, the underlying mechanism is the same: The lens, sculpted out of spacetime by a specific distribution of mass, deflects a beam of light. The geometry of the lens—its shape and size—dictates the effect it has on any light passing through it.

If we know the details of the lens—the precise curvature of spacetime induced by an object or mass distribution—we can predict the path that the light will take. In working out the prediction of general relativity that motivated the eclipse chasers, Einstein used the known mass of the Sun to calculate the path that light from a distant star would follow as it skimmed past the edge of the Sun.

More commonly, however, it is the lens itself that we are seeking. The power of gravitational lensing as a cosmological tool is that it is sensitive only to mass—dark matter, normal matter, matter that emits a blaze of light, matter that barely glows, or matter that is completely dark are all one and the same. It is only the amount of matter—the mass—and its distribution that determine the geometry of the lens. Einstein's Telescope offers the ability to image the mass content of the Universe.

Because most of the mass in the Universe is dark, most of the gravitational lenses are dark, or have a large dark component. A galaxy is essentially a large sphere of dark matter with a smaller visible collection of stars at its center, and most of a galaxy's power to act as a gravitational lens comes from its dark matter component. A black hole is not directly visible in any telescope, and the only hope of spotting this extreme incarnation of a warped spacetime is via its gravitational shenanigans.

Gravitational lensing allows us to determine the mass of all the matter in a galaxy or to detect the presence of a black hole by looking for the imprint it creates on the spacetime around it. Each lens creates a signature image: the curves and bumps in spacetime that define the lens redirect light from a distant source as it makes its way to our more mundane telescopes made of metal and glass. We can then work backward from this image, deducing the shape and size of the lens that must have created it.

A LIGHT-BENDING EXPERIENCE

The geometric fundamentals of gravitational lensing are very similar to the geometric optics of a lens ground from glass or plastic—in fact, it is possible to machine a lens out of plastic that reproduces the effects of a simple gravitational lens. Or you can conduct a simple experiment yourself by looking through the end of an (empty) wineglass. The shape of the stem as it curves into the base of the wineglass mimics the curve of spacetime, and you can replicate some of the effects discussed in this chapter.

The underlying physics is, of course, very different for glass versus gravitational lenses. In the case of an optical lens, light is bent when it travels from one material into another (from air into glass, for example). The curvature of the interface between these two materials and the different speeds with which light propagates in each medium (the index of refraction) dictate the redirection of the light. The amount of bending in a lens constructed of glass also depends on the color, or wavelength, of the light.

Light travels at different speeds in different materials—in general, it slows down in comparison with the speed of light in a vacuum—because the light interacts with the electrons in the atoms that comprise the material. Somewhat like a politician moving through a crowded room, stopping to shake hands along the way. The interactions slow the passage of the light (or politician) through the material. The net effect of this slowing is summarized in the index of refraction for a given material—the higher the index of refraction, the slower the light speed. The index of refraction also depends on the color of the light. In glass, for example, blue light travels more slowly than red light. The more light is slowed, the larger the angle through which it is bent. A beam of white light, which contains all colors, can thus be spread out into a rainbow of colors as it passes through a glass prism, each color traveling through the glass at a slightly different speed and bending accordingly.

The effects of gravitational lensing are due to the curvature of spacetime. It is not a change in the material that the light passes through that causes the bending, but a change in the strength of the gravitational field. Light does slow down in the presence of gravity, but this effect is completely independent of the color of the light. All colors are bent by exactly the same angle. This angle depends on how much spacetime has been warped and the incident path of the light—the closer it comes to the center of the dimple in spacetime, the larger the deflection.

The simplest lens consists of a single massive object whose size is negligible compared to the distances involved. The image produced by such a lens is determined by the mass of the object that acts as the lens of the telescope (how large a dimple in spacetime it creates), how closely aligned the lens and source are (the strongest lensing comes from a lens directly in front of the source), and the distances between the source, lens, and observer (lensing is strongest when the lens is exactly halfway between the source and the observer).

At the heart of all gravitational lensing, no matter how complicated, is the basic process shown in Illustration 4.4. As light travels in the vicinity of an object with mass, it follows the shortest possible path in the curved spacetime around the object. To understand this simplest case, we assume that this object is the only thing in the Universe (or at least there is nothing else nearby), so that far from the object spacetime is essentially flat and the light will travel in a straight line. The amount of bending is the angle between the original direction of the

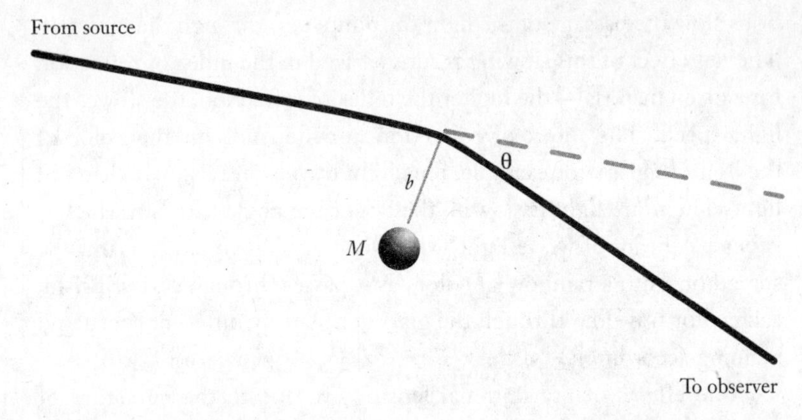

ILLUSTRATION 4.4 Gravitational lensing. The path of light from a distant source is deflected by an angle Θ when it passes by a mass *M*, on a path whose point of closest approach is the distance *b* (the impact parameter).

light (far from the massive object) and the direction in which it travels after it is far past the mass. This angle is calculated from Einstein's equations[13] and depends only on the mass of the object and the distance of closest approach between the object and the path of the light (this distance is called the *impact parameter*).

The bending angle Θ can be written very simply as the mass of the object *M* (multiplied by Newton's constant *G* to get the strength of gravity right) divided by the impact parameter *b* (multiplied by the speed of light squared to make all the units work out):

$$\Theta = 4GM/bc^2$$

The deflection angle increases as the mass of the lens increases. Light whose original path comes closer to the lens will also be bent more dramatically.

The image we then see here on Earth is a bit of an optical illusion: the light will appear to be coming from a spot on the sky that is shifted from where it would be in the absence of the intervening massive object. Just as the stars in Sir Eddington's photos were shifted from their normal positions in the cosmic tableau.

OPTICAL ILLUSIONS

One of the most striking effects of gravitational lensing is the creation of an Einstein ring, which occurs when the light source, the lens, and the observer line up exactly, with the lens more or less halfway between the source and the observer. This rare lensing phenomenon was first observed by Jacqueline Hewitt and her collaborators in 1987, when they discovered an Einstein ring in radio wavelength observations of quasars.[14]

Eight new Einstein rings joined the rogue's gallery of exotic gravitational lenses in 2006 (see Color Illustration 1) courtesy of a joint

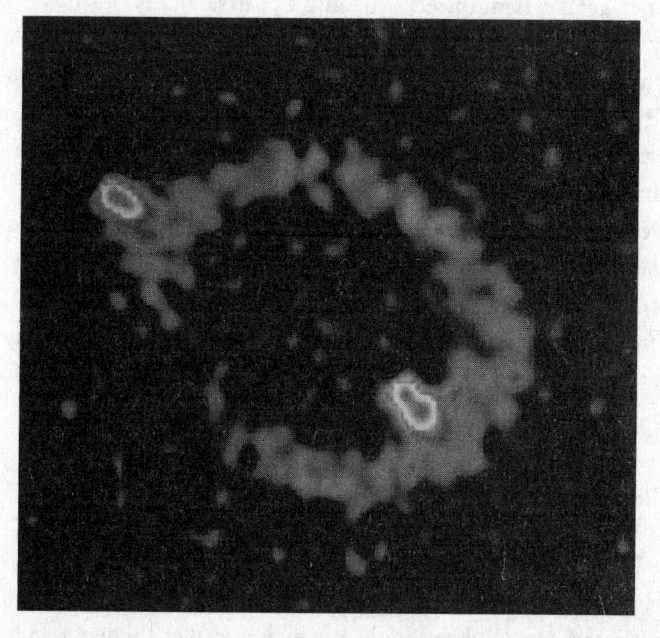

ILLUSTRATION 4.5 The first Einstein Ring. In 1987, Jacqueline Hewitt and her collaborators discovered the first Einstein Ring using the Very Large Array (VLA) radio telescope in New Mexico. The distant source is a galaxy that shines brightly in radio light, with a central core and two lobes of radio light extending out from either side of this core. The mass of an intervening galaxy (which can't be seen in this image, because it doesn't emit radio light) lies directly over one of the radio lobes and acts as a lens to produce two images of the central core (at about five and eleven o'clock) and a ring of light, created by lensing of one of the radio lobes.

project between the Sloan Digital Sky Survey (SDSS)—an ambitious observational tour de force that has collected and cataloged millions of galaxies, quasars, and stars—and the Hubble Space Telescope. Deliberately sifting through its data for images of gravitational lensing, the SDSS identified lensing candidates, which were then observed again with the Hubble's keener vision. The bull's-eye patterns revealed in the Hubble images are Einstein rings—an optical illusion produced by a warp in spacetime.

In each one of these images, the lens is an elliptical galaxy at a distance of between 2 billion and 4 billion light-years from Earth, and visible at the center of the bull's-eye as the large, bright yellow-orange ellipse of light. The bluish ring of light encircling the galaxy is actually another galaxy, lying directly behind the first, but about twice as far away from us. If the closer galaxy were not there, the more distant galaxy would appear as a small blue smudge of light.

This outlandish distortion of the original image illustrates one of the stranger implications of Einstein's curved spacetime: once you add lumps to spacetime, there is more than one way for light to get from point A to point B. In flat (empty) space, light that starts out in a particular direction keeps heading that way. There is only one path between two points.

In a curved spacetime, however, light from a quasar or galaxy that starts out in two (or more) initially divergent directions can follow paths that curve around and ultimately end up in the same place. Further, gravitational lenses are in some ways maximally aberrant. Unlike an optical lens, which is designed to reproduce the original object as faithfully as possible at a particular focal point, each part of a gravitational lens has its own focal point. The lens sends the light that passes through it off in different directions, depending on which part of the lens it intercepts (and the angle at which it strikes the lens). What we see from our particular position on Earth is all of (and only) the light that has been redirected to where we are sitting.

Einstein rings are created as shown in Illustration 4.6a. In the absence of an intervening lens, the light would travel from the distant galaxy to us in a straight line. When the lens sits in front of the

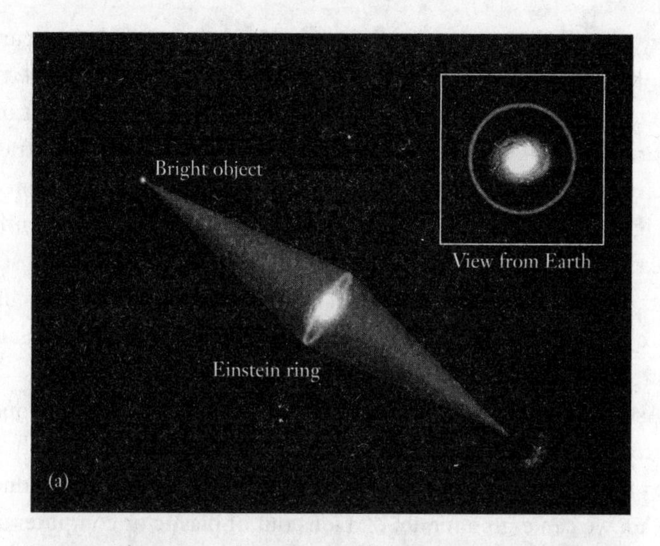

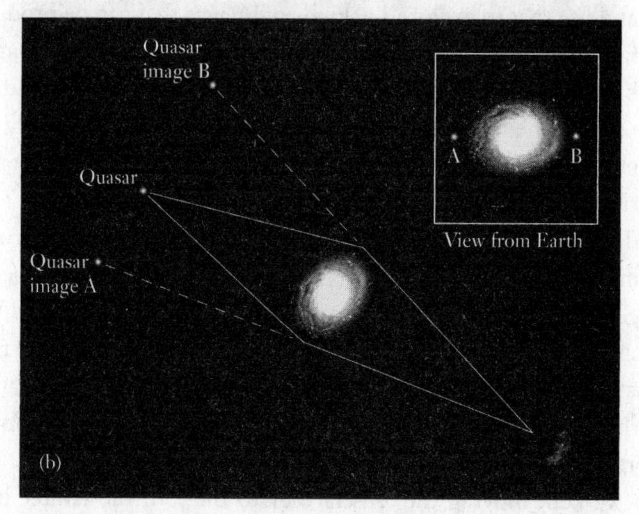

ILLUSTRATION 4.6 An artist's visualization of gravitational lensing. (a) A distant source of light such as a galaxy lies directly behind a closer galaxy. The light from the more distant source is deflected by the lensing power of the mass of the closer galaxy, bending into a ring. The inset box depicts what would be seen through the Hubble Space Telescope, for example. (b) If the distant source (in this example a quasar) does not lie directly behind the lensing galaxy, two images of the quasar are produced, as seen in the inset box.

galaxy, the light that eventually reaches us has traveled on a curved path, bending around the center of the lens. But there is more than one curved path. Light that intercepts the lens a bit to the left of its center is bent from its original path to one that now points straight to Earth. But so does light that strikes the lens an equal distance to the right of its center. Or the same distance above or below. In fact, since there is no left, right, up, or down in space, the preceding statements are a bit ambiguous—which is why we see a ring of light. There is an entire ring of possible paths from the distant galaxy, through the lens, to the Earth.

Astronomical objects are great for observing, but not so cooperative for conducting experiments—we can't move a lens (such as a galaxy) around in space to see what kinds of images can be produced. Instead, we can craft a model of a lens out of plastic or computer software and use that to explore the kinds of images that can be produced. Illustration 4.7a is the original, unlensed image of a question mark on a dark background, representing a distant source of light. The next frame shows the effect of placing a "gravitational lens" directly over the question mark. The result is a ring of light, in which the details of the original question mark are hard to distinguish.

The lens was moved just slightly to the left of center to produce the image in Illustration 4.7c. The Einstein ring has disappeared, replaced by two separate images of the question mark, one of which is smaller, fainter, and upside down. The image in Illustration 4.6b traces the path of the light in this case. The symmetry of the first example, with the lens directly over the question mark, has been broken, and the light reaches us from two directions on the sky.

If we kept moving the lens farther to the left, the fainter image would get smaller and fainter, until it eventually disappeared completely, leaving only a single image. This image would still be slightly shifted away from the lens, and slightly brighter, but both of these effects would also disappear as the lens moved farther away. Eventually, we would see only the original question mark again.

The dividing line (for a simple galaxy lens) between one and two images is a circle—an imaginary line drawn around the lens known as

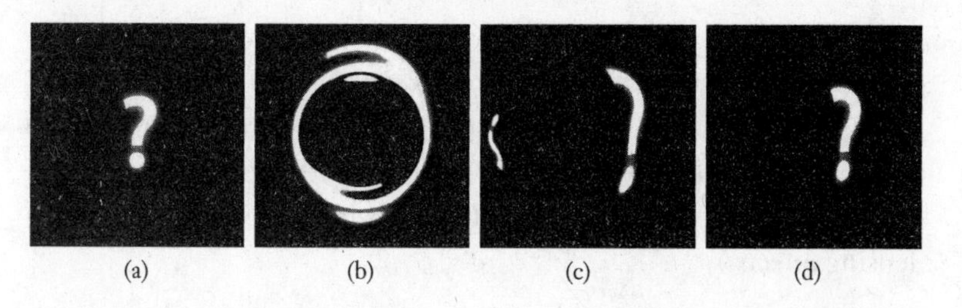

ILLUSTRATION 4.7 The effects of gravitational lensing, illustrated with a computer program to simulate the bending of light by a massive (invisible) object between you and an image of a question mark. (a) Original, unlensed image of a question mark. (b) Here the lens lies directly over the question mark, and a ring of light known as an Einstein ring is produced. (c) The lens in this frame is just to the left of the question mark, producing two images—note that one is inverted. (d) The invisible lens is at the left edge of the frame, and only weak lensing effects are seen. The image of the question mark is slightly distorted and shifted a tiny bit to the right.

the *Einstein radius*. A distant source of light such as a quasar or galaxy that lies within this circle will be lensed into two images: one that is smaller, fainter, and flipped, and a second that is a magnified version of the original. If the light source lies outside of the circle, only one image is seen (Illustration 4.7d). And if the source lies directly behind the lens, the Einstein ring that is produced traces out the Einstein radius—a beautiful visualization of a mathematical equation.

THE SCALE OF THE PROBLEM

The Einstein radius thus sets a natural scale for lensing. It is, in a sense, a way of characterizing the effective reach of a gravitational lens. The size of the Einstein radius depends on the size and depth of the dimple in spacetime created by the lens—the more massive the lens, the larger its corresponding Einstein radius. It also depends on where the lens lies along the line of sight to the source. The Einstein radius is largest when the lens lies halfway between the source and the observer, and its size decreases as the location of the lens moves closer to (or farther from) the source than this optimum point. This is a focusing effect, analogous to adjusting the focus on a telescope or pair of binoculars by moving a lens back or forth in order to get the correct focal length for your eyes. A gravitational lens is most powerful when the lens is located at the halfway point between observer and source.

The Einstein radius also sets the size of an Einstein ring or the maximum separation of multiple images, and by using the Einstein radius as a guide, we can visualize the probability of seeing interesting lensing effects.

MULTIPLYING, MAGNIFYING, AND A BIT OF A STRETCH

A point of light (such as a star or quasar) viewed through a gravitational lens that is very small compared to its Einstein radius can appear as a magnified image, two images, or an Einstein ring as described in the preceding discussion. But if the lens or the source of light—or both—is more interesting, so are the images that can be created.

The first complication to add to the mix is to consider a lens that is not perfectly round. An elliptical galaxy has both a size and a shape that sculpt a very different lens from the simplest case. It does not have the spherical symmetry of the previous example—an ellipse has a long axis and a shorter one—so its orientation adds another dimension to the lensing equations. The images that such a lens produces depend not only on the relative positions of lens, source, and observer, as before, but also on the directions in which the axes of the ellipse line up. A single quasar that lies almost directly behind an elliptical galaxy can be focused by the lens into four separate images[15]—it appears as a quartet of (identical) quasars arranged in a pattern around the lensing galaxy known as an *Einstein cross* (see Illustration 4.8).

Next, add some structure to light sources. Quasars are essentially points of light, and points can be magnified and multiplied, but having no shape (by definition), they cannot be otherwise distorted. Galaxies, on the other hand, come in a variety of sizes and shapes. If a distant galaxy lines up almost behind a foreground lens, it can be multiplied, magnified—and stretched. The question mark in our toy model of lensing illustrates this beautifully. The images that are produced when the question mark is lensed are not exact copies of the original, but

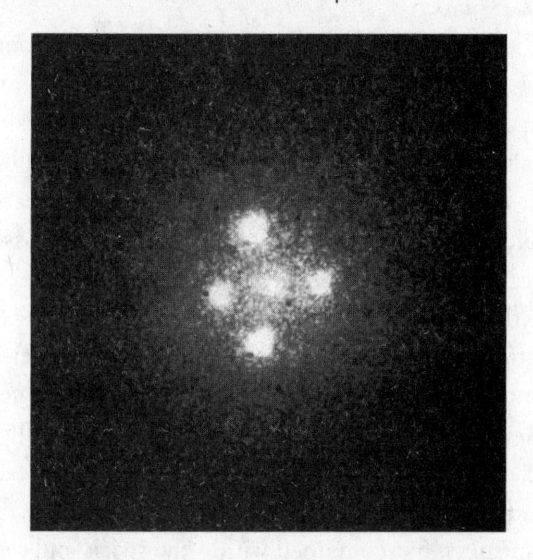

ILLUSTRATION 4.8 The Einstein cross. In this image taken in 1990 by the European Space Agency's Faint Object Camera on board the Hubble Space Telescope, four copies of a single distant quasar (about 8 billion light-years away) are visible in a crosslike pattern around the central region (bright center spot) of a galaxy that is much closer to us (roughly 400 million light-years from Earth). The mass of the galaxy acts as a lens to bend the light from the quasar into the four copies seen.

stretched and curved around the center of the lens. If the Einstein ring were drawn on the illustration, the twin images would lie close to the ring—one just inside, the other just outside of it, and both curved to follow the arc of the ring.

Color Illustration 2 provides one of the most extreme examples of this distortion. The bright red-orange object in the center of the image is a large elliptical galaxy about 6 billion light-years away that is acting as a gravitational lens for an even more distant galaxy, which lies almost directly behind the lens at a distance of 10 billion light-years from Earth. The light from the distant galaxy has been stretched into the long blue arc that is seen just above the center of the lensing galaxy, its original shape distorted as it passed through the warp in spacetime produced by the massive foreground galaxy.

Clusters of galaxies offer an even more psychedelic window into our curved spacetime. These conglomerations of hundreds of galaxies, embedded in a huge sphere of dark matter that far outweighs the combined weight of the individual galaxies within it, are powerful lenses with masses thousands of times greater than a single galaxy. They are also much more complicated lenses. The sphere of dark matter that comprises the bulk of the cluster mass creates an enormous dimple in spacetime, with additional lumpiness tossed in by the galaxies within the cluster. Seen through this multifaceted lens, the light from the background field of galaxies weaves its way through the undulations imposed on spacetime by the cluster, emerging in a riot of giant arcs and smaller *arclets* that encircle the center of the cluster, tracing out the mass of the lens (see Color Illustrations 4 through 7).

The Van Gogh–like appearance of these Hubble Space Telescope images is uncanny, but contained within the swirls and replications is the imprint of the lens—its size, shape, and mass distribution. Five copies of a single blue galaxy with distinctive markings (the dark patches are created by clouds of dust within the galaxy, which block the starlight in this young, star-forming galaxy) can be seen in Color Illustration 5: three along the upper left side of the image, one opposite these in the lower right corner, and one at the center. From the size, position, and multiplicity of these arcs, as well as the many arclets of other background galaxies that also appear in this image, the lens can be reconstructed—and its hidden trove of dark matter revealed.

THE ART OF LENSING

Therein lies the true power of lensing—it allows us to trace out the matter in the Universe and to probe the structure of spacetime itself. The distortion caused by this matter can be extremely subtle or outlandishly obvious, depending on the degree to which spacetime has been deformed and on our ability to detect the resulting image. The trick is to learn how to translate the pattern of images we see into the size and shape of the lens that created it.

Using Einstein's Telescope, we are enthusiastically expanding our horizons in previously unimaginable ways, from the search for dark objects within our own Galaxy to the quest for the nature of the dark energy that controls the fate of the Universe. Our explorations of the cosmos and the microcosmos have led us to a new mystery—and a new forensic technology with which to solve it.

CHAPTER 5

MACHOs and WIMPs

The search for dark matter begins at home—in our own Galaxy. Galaxies, including the Milky Way, are overwhelmingly dark. Their visible matter is revealed in some of the most gorgeous views of the Universe ever seen. Majestic spiral arms, brilliant central cores, dark rings of gas and dust that define the plane of the disk. Fiercely lit by the output of perhaps hundreds of billions of stars, with the starlight reflecting off the gas and dust from which the stars were formed and to which they will return, galaxies are the luminous tracers with which astronomers map the structure of the cosmos.

However, these magnificent beacons are truly the tip of the iceberg for cosmological navigators. The visible components of a galaxy—stars, gas, and dust—viewed by any telescope sensitive to light of any wavelength, represent only about one-tenth of the entire galaxy. The bulk of a galaxy—about 90%—is in the form of dark matter.

THE MILKY WAY

Looking up at a dark night sky, far from any city or other source of light pollution, it appears as though a beautiful white band had been painted with a broad brush against the inky darkness. It is an unfortunate consequence of modern life in the United States that many people never have the opportunity to see this incredible edge-on view of our home Galaxy, the Milky Way. In addition to the band of the Milky

ILLUSTRATION 5.1 The Milky Way. The disk and central bulge of our home Galaxy are beautifully displayed in this edge-on view, taken by the DIRBE (Diffuse Infrared Background Experiment) instrument on board the COBE satellite.

Way, thousands of stars are sprinkled across the dome of the sky. All of the stars that can be seen with the naked eye, even from the most advantageous location, are part of our Galaxy and, in fact, are from a very small, nearby region of the Galaxy.

Our home Galaxy is now known to be one of hundreds of billions of galaxies in the observable Universe, and part of a small local group of galaxies. Within this local group our largest neighbor, the Andromeda galaxy, is zooming toward us from a distance of about 2 million light-years away, on a course that may end in a spectacular collision with the Milky Way roughly 5 billion years from now. Two smaller members of the local group are the Large and Small Magellanic Clouds (LMC and SMC), galaxies that are circling the Milky Way at a distance of about 170,000 light-years, caught in the gravitational pull of our own Galaxy like satellites orbiting the Earth.

The Milky Way is a fairly typical spiral galaxy, with a Frisbee-like disk of stars, gas, and dust roughly 80,000 light-years from one side to the other, and a few thousand light-years thick. In the very center is a dense core of stars surrounding a central black hole whose mass is more than 2 million times that of the Sun. The band

ILLUSTRATION 5.2 A spiral galaxy. Known as M101 and located at a distance of 25 million light-years in the direction of the constellation Ursa Major, this spiral galaxy is seen face on and shows the pattern of spiral arms and central region of high stellar density that is also found in the Milky Way (a similar image of the Milky Way could be obtained only from far outside our home Galaxy).

of the Milky Way seen from Earth is an edge-on view of the disk as we look toward the center of the Galaxy from our location near the edge of the disk, about 20,000 light-years from the center. A view looking down on the disk, which is not possible to obtain since one would have to travel (or send a telescope) tens of thousands of light-years away, would reveal an image similar to that obtained of other spiral galaxies, with graceful spiral arms pinwheeling about the central core.

However, even this incredible view would not reveal the vast bulk of the Galaxy. In fact, a single image of the entire Galaxy would not even hint that anything more was there. But a video, a sort of time-lapse photography spanning hundreds of millions of years, would send up an astronomical red flag, making it clear that something—a lot of something—was missing. The stars would be seen circling about the center of the Galaxy at speeds far beyond the limit imposed by the gravity of the visible matter.

WEIGHING THE GALAXY

To truly understand and describe our home Galaxy, we need to determine how big it is and how much stuff is in it. In other words, how much does it weigh? We obviously cannot plunk the Galaxy down on one side of an enormous balance scale in order to weigh it. Instead, astronomy has several ways of measuring the mass of an object in space.

The first method is essentially an exercise in addition. Given the mass of the smaller objects (like stars) that make up a larger system (a galaxy), counting the number of the smaller objects and adding all of their masses together should yield the total mass in the system. For our Galaxy, this means adding up the masses of all the stars and the gas and dust that make up the visible part of the Galaxy. In doing so, we find a grand total of about 60 billion times the mass of the Sun.

How do astronomers know how much mass is in each star? Determining a star's mass requires a second method: observing how things move in space. The Universe is incredibly dynamic—an active and often violent arena—where nothing is ever fixed in time or place. Moons whirl around planets, while the planets orbit their home stars; stars orbit about the center of their galaxy, galaxies dance around each other in small groups or large clusters, and space itself accelerates in a great cosmic stretch.

Gravity is the main force responsible for the motions of moons, planets, and stars. Because we understand how gravity works (at least over distances of astronomical size), we know that the pull that objects exert on one another depends on how much mass they have. Using the observed motions of objects in space, we can calculate the mass that gives rise to their movements.

The gravitational pull of the Sun's mass is tugging inward on each of the planets in the Solar System, pulling them in toward the Sun. This inward tug must be exactly counterbalanced by the velocity (speed and direction of motion) of each planet if it is to remain in orbit about the Sun. The closer a planet is to the Sun, the stronger is the pull of the Sun's gravity, and the faster the planet must be moving to maintain its orbit. An object moving too slowly will be pulled into a fiery plunge into the Sun; one moving too fast will zoom past the Sun, forever escaping its gravitational clutches.

We can use this understanding of gravity to calculate the mass of the Sun—essentially weighing it. Given the speed with which a planet orbits the Sun, and its distance from the Sun, we can determine the Sun's mass. If, for example, the Sun were a more massive star, a year on Earth would be less than 365 days. The Earth would need to circle a heavier Sun faster in order to maintain its orbit at the same distance from the Sun. Likewise, we expect a planet closer to the Sun to be moving more quickly, and thus have a shorter year.

When we look at all of the planets, we find that their velocities and distances all give the same mass for the Sun (1.989×10^{30} kilograms), and that the farther a planet is from the Sun, the longer a year on that planet is—exactly as predicted. The length of a year on Mercury is about 88 Earth days; on Mars, a year is 687 Earth days; and on distant Pluto, a year is a whopping 247.7 Earth *years*.

To measure the mass of a star, a similar method is used. Astronomers first look for stars that are part of a pair of stars. The motions of such stars as they orbit around each other reveal their masses. From observing many pairs of stars of different types—different color and brightness—astronomers now have a reasonably accurate knowledge of the mass of a star of a given type.

Can we use a similar method to weigh the galaxy? Yes. In a galaxy, most of the stars are in the disk,[1] and these stars circle around the center of the galaxy just as the planets orbit about the Sun—although the timescales are far longer. The Earth takes one year to complete its orbit about the Sun, while the Sun takes about 250 million years to orbit the center of the Galaxy, trucking along at 200 kilometers per second (almost half a million miles per hour). However, the same laws of gravity that keep the Earth and other planets in orbit about the Sun are responsible for the movement of the stars in a galaxy about its center.

Of course, the stars are part of the galaxy, not outside of it, which

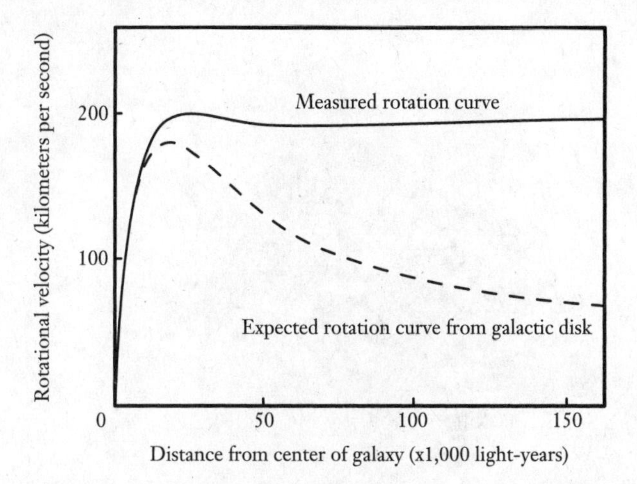

ILLUSTRATION 5.3 The galaxy rotation curve. The speed of stars and gas about the center of a typical spiral galaxy as a function of their distance from the center (solid line). Stars 100,000 light-years from the center are orbiting as fast as those much closer in, and much faster than expected from the visible mass in the galaxy (dashed line). Such "rotation curves" have been measured for over 1,000 galaxies, including the Milky Way, and all exhibit a similar pattern.

complicates things a bit. The speed with which a star travels through a galaxy depends not on the total mass of the galaxy, but on just the part of the galaxy that is contained inside of the star's orbit—that is, all of the mass closer to the center of the galaxy than the star is. Thus, the motion of a star close to the center tells us how much mass is in the central part of the galaxy; a star farther out near the edge of the disk reveals how much mass is within the disk region of the galaxy. Ideally, we want to find objects completely outside of the galaxy in order to measure its total mass. For this purpose, astronomers turn to star clusters—groups of 100,000 or so stars—that are found far outside of the galactic disk. Star clusters act as galactic satellites, orbiting the galaxy at distances of hundreds of thousands of light-years from the center. The orbits of these star clusters reflect the gravitational pull of the galaxy—and thus reveal its mass.

The combined observations of the stars in the disk and the more distant star clusters present a very different pattern from what was expected. Their motions make no sense if all of the mass of the Galaxy is in the stars and gas that we see. Just as the speed of the planets decreases

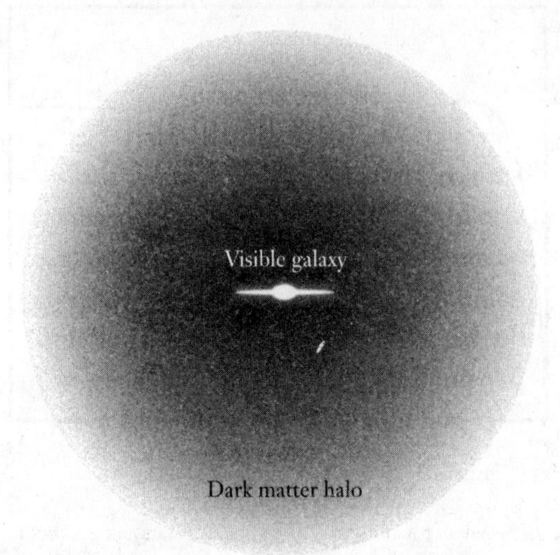

ILLUSTRATION 5.4 An artist's rendering of the dark matter halo surrounding the visible part of a galaxy. Seen at the center is an edge-on view of a spiral galaxy (similar to the image of the Milky Way in illustration 5.1). The galactic disk and central bulge are immersed in an enormous sphere of dark matter, known as the galactic halo, which extends far beyond the visible components of the galaxy.

with distance from the Sun, the laws of gravity predict that the farther away an object is from the center of the Galaxy, the slower its speed should be. Instead, we find that as we move out from the center of the Galaxy, the stars and star clusters are moving much faster than predicted. Stars farther from the center than the Sun is are still moving as fast as the Sun, when they should be orbiting more slowly. This means that unless there is more gravity—more mass—to keep these stars in orbit than we can account for by adding up the mass in the stars and gas in the galactic disk, they should have been flung out of the Galaxy long ago. Their speeds are exceeding the "escape velocity" of the Galaxy.

THE DARK SIDE OF THE GALAXY

This egregious flouting of the laws of gravity has now been measured for over 1,000 galaxies in addition to the Milky Way, and in each galaxy the velocities insist on the same conclusion: there is not enough mass—by a huge factor—to account for the velocities that are observed.

What is incredible is that, in order for the velocities to make sense, an enormous amount of matter must be added to each galaxy. Over 90% of the mass of a galaxy must be in the form of matter that is not seen by our telescopes. Dark matter. This dark matter permeates and surrounds the disk of the galaxy, in a huge sphere that extends out over half a million light-years from the center of the galaxy. This sphere of dark matter is known as the *galactic halo*.

What is this mysterious dark matter? What do we know about it? What is it made of? Can we ever see it? Capture it? Cook up a batch of it in the laboratory? The only evidence we have so far that dark matter even exists comes from its gravitational effects on objects that we can see. Dark matter is exactly that—dark. It is either (1) too dim to be seen, even through a telescope, or (2) made of some substance that does not emit, absorb, or reflect light.

These two possibilities delineate the two main categories of dark matter. The first category contains things that we already know exist, such as planets, faint stars, or black holes, and goes under the heading of *MACHOs*. The second category comprises the unknown—entirely new kinds of particles that are collectively known as *WIMPs*.

This second type of dark matter is best summed up by the immortal words of Monty Python: "And now for something completely different." Particle physics (the science of the smallest objects in the Universe) and astronomy (the science of the largest objects in the Universe) have joined forces to propose that the dark matter is a new *kind* of matter. This dark matter would be very different from normal matter—the stuff of which people, planets, and stars are made. It would be something even stranger than antimatter. The most likely candidate for dark matter (so far) is a particle that has never yet been seen—a particle that scientists have christened the WIMP.

WIMP is an acronym standing for *weakly interacting massive particle*. WIMPs barely interact with normal matter, streaming through the Earth, for example, almost as if it were not there at all. This makes them very hard to detect and impossible to "see" in the sense that they don't emit or reflect light. Scientists don't yet know exactly how heavy these particles are, but expect that they are roughly a hundred times heavier than a proton. This is still only the tiniest fraction of a gram, but relatively heavy for a basic particle.

More important, scientists don't know exactly *what* WIMPs are. If WIMPs exist, they play an extremely important role in the evolution of the Universe and the structure within it, and we will return to them later in this book. But before searching for something that has never been seen, it makes sense to look first for objects already known to exist, with the hope that we may have underestimated their numbers[2]—to look for **ma**ssive (in the more usual sense of the word, i.e., as heavy as the Sun), **c**ompact (like a star, as opposed to a diffuse cloud of particles) **h**alo (located in the halo of our Galaxy) **o**bjects, or MACHOs.

This somewhat abominable, if memorable, terminology was initially suggested as a joke over lunch. WIMPs had been part of the scientific lexicon for some time when experiments to detect dark matter in the form of planets and dim stars were first being discussed. The battle lines for the type of dark matter that would, in a very real sense, rule the Galaxy were being drawn. "Was the Galaxy filled with WIMPs? Or would less exotic candidates for the dark matter be found? Would it be WIMPs or . . . MACHOs?" As soon as the name "MACHO" was spoken, the speaker (Kim Griest, currently a professor of physics at the University of California at San Diego) tried to retract it. But it was, of course, too late.

MACHOS

MACHO—massive compact halo object—is a generic name representing a motley mix of candidates. The initial list of contenders included dim stars, black holes, and Jupiter-sized planets. Most stars—those like the Sun, for example—are far too bright to be MACHO candidates. We would see them if there were enough of them in the Galactic halo to account for the missing dark mass. Several types of stars are very dim, however, and such faint stars could escape detection with even our most powerful telescopes.

All of these dim stars are the end product of some part of the general cycle of stellar birth and death. A star forms out of a dense cloud of gas (mainly hydrogen with a sprinkling of other elements) that collapses under the weight of its own mass. If there is enough mass in the forming star, gravity pulls it together so strongly that the density and temperature in the very center reach incredibly high values—high enough to ignite nuclear reactions. The star lights up and, like the Sun, continues to shine brightly as long as it has enough hydrogen gas to fuel the nuclear processes. A star like the Sun, fusing hydrogen nuclei to produce the light that provides the energy for life on Earth, will burn through its fuel in about 10 billion years. Larger stars are gas hogs and can tear through their fuel supply in much less time; smaller stars are much more conservative and may live a trillion years or more.

What happens once a star runs out of fuel also depends on how massive it is. The most massive stars, in keeping with their flagrant lifestyle, go out in a blaze of glory, exploding in a brilliant finale known as a *supernova*. The debris from this explosion is recycled back into the surrounding gas and dust, leaving only a small fraction of the original mass of the star compacted into a dense, dark remnant such as a black hole.[3] These black holes have masses a few times that of the Sun or higher, and they are the end game of the most massive stars—those that start out 10 to 100 times more massive than the Sun.

Stars that are initially only a few times heavier than the Sun also explode in supernovae when their fuel runs out, leaving behind a *neutron star* instead of a black hole. Composed of the densest state of matter known, a neutron star contains the mass of one and a half Suns packed into a sphere less than 10 miles in diameter.

Both neutron stars and black holes in the Galactic halo would be faint enough to escape detection by telescopes. Neither are true stars, in the sense that they are not actively burning hydrogen (or helium) to fuel the prodigious output of light that we commonly associate with stars. Rather they are stellar remnants, the fossil remains of massive stars.

Less massive stars, such as the Sun, also leave behind a fossil record in the form of a burned-out core. The death throes of the Sun and stars like it may not be as violent as those of their heavier counterparts, but they are spectacular in their own way (and, unfortunately for whatever inhabitants of Earth may exist roughly 5 billion years from now, spectacularly fatal for life on their orbiting planets). The Sun will eventually blow off its outer layers, and the inner core will collapse into a *white dwarf*, a dense stellar cinder with about half the mass of the Sun. A white dwarf will initially glow with great heat, then quietly cool and fade away over billions of years. Stars like the Sun are more numerous than their more massive sisters, and there have been many generations of stars since the light of the first stars pierced the darkness of the Universe 12 or 13 billion years ago. The remnants of these stars—white dwarfs—are difficult to detect, and their numbers are still being debated.

At the other end of the stellar life cycle are star wannabes. *Brown dwarf* stars are formed in a process similar to that which forms more massive stars, but in this case the collapsing cloud of gas does not have enough mass to light up as a full-fledged star. The mass of a brown dwarf is less than a tenth the mass of the Sun, and the gravitational pull of this mass on itself is not strong enough to ignite the nuclear processes. Brown dwarfs do not shine like other stars, but glow only very faintly as they give off a small amount of heat. Thus, they are extremely hard to detect—the first image of a brown dwarf was not obtained until 1995 by the Hubble Space Telescope. Such faint objects make excellent MACHO candidates, and although our current models of star formation predict far too few brown dwarfs to provide all of the dark matter, maybe there is another mechanism for populating the Galaxy with these low mass stars.

The final candidates for MACHOs are also the smallest—Jupiter-sized planets, with masses about 1,000 times smaller than the Sun. There has been much excitement in the past 10 years as astronomers have detected more and more planets outside of our Solar System. We know such planets exist, even if we don't yet know exactly how many of them there are in the Galaxy. However, there would need to be many more such planets than we expect from the number detected so far and from the theory of planet formation. And most of them would need to be free-floating in the Galaxy, not attached to any star.[4]

Black holes, neutron stars, white dwarfs, brown dwarfs, and large planets are all known to exist—a definite plus in the dark matter quest—and have at one time or another been considered as MACHO candidates. Whatever MACHOs might be, however, the more pressing question is whether they are indeed the dark matter in our Galaxy. Do MACHOs dominate the Galactic halo, or is the Galaxy full of WIMPs? And how does one look for MACHOs, which are impossible to see with our current telescopes?

TWINKLE, TWINKLE, LITTLE STAR

MACHOs may not be visible, but their mass will warp spacetime, deflecting light from distant stars as it travels past them. Fifty years after Einstein dismissed the possibility of observing this deflection, Bohdan Paczynski of Princeton University suggested a way to search for dark matter in the Galaxy using just this phenomenon. He outlined an experiment in which MACHOs act as gravitational lenses, focusing and amplifying the light from individual stars in a nearby galaxy, such as the Large Magellanic Cloud (LMC).[5]

The key ingredients in gravitational lensing are the mass of the lens; the distances between observer, lens, and light source; and how closely the lens and light source line up with each other. These ingredients are very different—by orders of magnitude—for lensing of stars by MACHOs than for lensing of quasars by an entire galaxy. The basic mechanism is the same, but the players are different. And it is this difference that makes the hunt for MACHOs so challenging.

In the examples of quasar lensing discussed in the previous chapter, the lens is an entire galaxy, which has a mass roughly a trillion times the mass of the Sun. Such a lens is powerful enough to produce the dramatic multiple images seen in the lensing of quasars. A typical MACHO mass, on the other hand, is much smaller—about half the mass of the Sun. Because a less massive object will cause a shallower warp in the surrounding spacetime and a correspondingly smaller deflection of light, gravitational lensing by a MACHO results in a much more subtle shift in the path of the light from a distant star. The star must be almost directly behind the MACHO in order for the star's

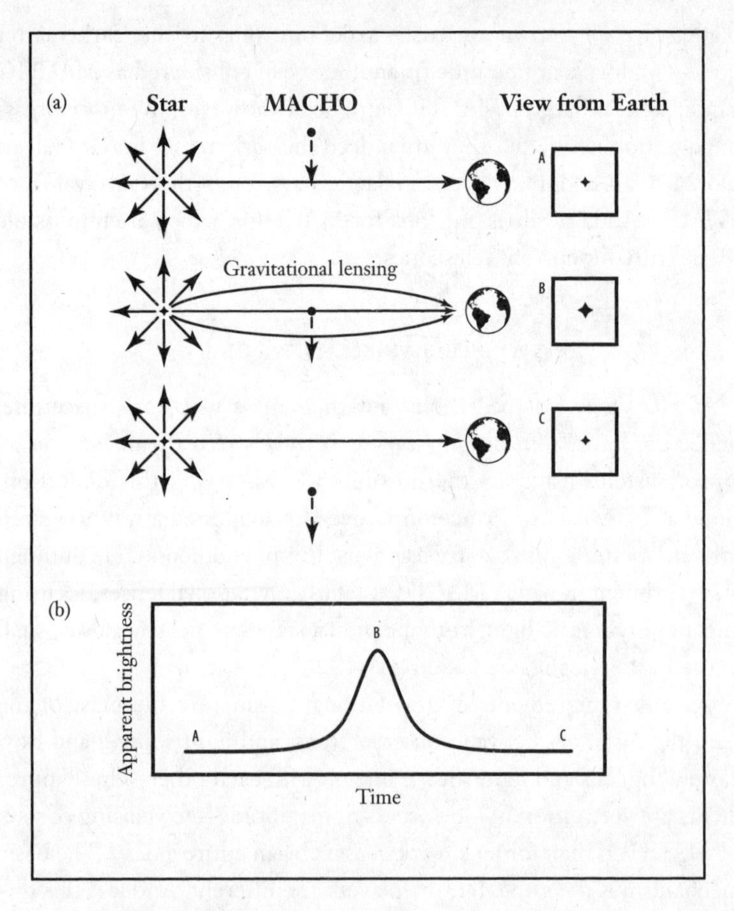

ILLUSTRATION 5.5 Finding MACHOs. MACHOs act as gravitational lenses when they move in front of a distant star. (a) Light from a star travels to a telescope on Earth, producing image A. As a MACHO passes in front of the star, it creates a dimple in spacetime that acts as a gravitational lens, and the star appears brighter (image B). As the MACHO continues to move past the star, the star's image will return to its original brightness (image C). (b) Astronomers track the apparent brightness of the star over time to produce a "light curve," as seen in this illustration of a typical MACHO lensing event.

light to be noticeably affected by the small dimple in spacetime caused by the MACHO.

Consider a typical MACHO candidate: a white dwarf with about half the mass of the Sun in the halo of our Galaxy roughly 30,000 light-years away from Earth. If this white dwarf is in front of a star that is 170,000 light-years away in the LMC, it will lens the light from the star and produce two images of the star. The separation of these two images will be only about a milli arc second (less than one-millionth of a degree) apart, and thus lensing by objects with masses smaller than a few times the mass of the Sun has been dubbed *microlensing*. This tiny angle of separation between the images will, as Einstein put it, "defy the resolving power of our instruments."[6] Even the most powerful telescopes of today are unable to see that there are two images. Instead, the light from the two images will add together and only a single image of the star will be seen—but it will be brighter.

Thus, searching for MACHOs means searching the LMC for stars that are brighter than they would be if they weren't being observed through a MACHO lens. The problem is that we don't know how bright a given star should be. Astronomers have a wealth of data on different types of stars—stars of different colors, for example—and in many cases they have an idea of roughly how intrinsically bright a star of a certain type is. However, even if the intrinsic luminosity of a star were known to high precision (which in most cases it isn't), the exact distance to the star must also be known in order to ascertain how bright the star should appear to an observer on Earth—and we don't have any good ways of determining this distance either.

MACHO searches would not be possible if it weren't for the fact that MACHOs, whatever they are, move in space in a way dictated by gravity. MACHOs in the halo of the Galaxy are traveling on orbits about the center of the Galaxy (although they are not confined to the disk as most of the stars are). The more mass there is in the Galaxy, the stronger the pull of gravity is on the MACHOs and the faster they speed through the halo. A typical MACHO may be moving at roughly 270 kilometers per second (almost 600,000 miles per hour).

As the MACHOs move through the halo, occasionally one passes in front of a star in the LMC. When this happens, the star is lensed. It appears brighter and brighter as the MACHO approaches, then fades back to its original brightness as the MACHO moves past.

The mass of the MACHO creates a little dimple in the fabric of spacetime that moves with it. The more massive the MACHO, the larger the dimple. This dimple is the effective lens, bending the light of the star and making it appear brighter. The size of the dimple is characterized by the Einstein radius of the lens—the more massive the MACHO, the larger the Einstein radius. The light from the distant star must pass close enough to the MACHO to be inside this radius in order for lensing to occur. The farther inside the Einstein radius the light passes—that is, the closer the MACHO comes to lining up directly in front of the star—the stronger the lensing will be and the brighter the star will appear.

One way to picture what is happening is to think of the MACHO as a magnifying glass, whose size (radius) is given by the Einstein radius. As the MACHO moves through space, it is as if a magnifying glass is being passed in front of the stars in the LMC. If the magnifying glass passes directly in front of a star, the star will be magnified and appear brighter. The power of this magnifying glass is stronger in the center, so when the star first falls behind the glass it is magnified only a little. The star appears ever brighter as the magnifying glass sweeps over it until it is nearest to being directly under the center of the glass, when it reaches a maximum brightness. As the magnifying glass continues to move over it, the star appears dimmer and dimmer until it returns to its original, unmagnified brightness.

All that astronomers need to do, then, is monitor the light from stars in the LMC and wait for one to get brighter, then dimmer again—in other words, to twinkle.

How long is the wait? This is where another challenge to detecting MACHOs comes in. In order for lensing to occur, the MACHO must pass almost directly in front of a star. To continue the magnifying-glass analogy, the magnifying glass is extremely tiny—much, much smaller than the typical distance between stars in the LMC. At any given time it is much more likely that the MACHO is in front of nothing than in front of a star in the LMC. And although the typical speed of a MACHO is enormous relative to speeds we usually experience here on Earth, it still allows the MACHO to cover only a small fraction of the distance between the LMC stars.

Consider an optimistic scenario in which we assume that all of the dark matter in the Galaxy is in the form of MACHOs, moving at typical speeds of about 270 kilometers per second. The odds that a star in the LMC will be lensed by one of these MACHOs during a year of watching is roughly one in a million—better than your chances of winning the big jackpot in the lottery, but still seemingly impossible.

Science has a classic and very effective method for dealing with such difficulties. If the probability that a star will be lensed is one in a million, then in watching a million stars, the odds are that you will see one of them lensed. So in order to be assured of seeing something (assuming enough of something is really out there), settle back in your comfy observatory chair and plan on monitoring millions of stars over a period of several years. Such a program would have been unrealistic and unrealizable for a lone astronomer in 1936 when Einstein wrote his paper on lensing. It was still an ambitious, but not impossible, undertaking in 1986, when Paczynski (who was also somewhat skeptical of the possibility of actually observing a microlensing event) outlined a potential experiment. However, with the aid of computers that are capable of handling large data sets and in the company of a large collaboration of scientists, it might just work.

THE SEARCH BEGINS

In spite of the daunting obstacles, the search for MACHOs began in earnest in 1990. Three teams of scientists from around the world—(1) the MACHO collaboration, a joint US/Australia effort led by Charles Alcock, then at Lawrence Livermore National Laboratory; (2) the EROS collaboration (Expérience pour la Recherche d'Objets Sombres), a team of French scientists headed by Michel Spiro; and (3) the OGLE collaboration (Optical Gravitational Lensing Experiment), a joint US/Poland experiment under the leadership of Andrzej Udalski—traveled to the Southern Hemisphere in order to observe stars in the Small and Large Magellanic Clouds.[7] These two satellite galaxies, our nearest galactic neighbors, are visible from the Southern Hemisphere, and they can be seen with the naked eye as irregular cloudy patches.

All three of these experiments used a similar approach, although the details of their observing and analysis techniques were different. They chose certain fields of stars in the LMC to serve as the light sources. They observed these stars on a regular basis (typically imaging the same field of stars several times a week, depending on the weather and the full moon) for several years, and carefully compared the brightness of each star to its brightness in previous images. This comparison was done with computer software that took into account the fact that the conditions under which each image was taken were different—on some nights the sky was clearer than others, for example, so that all of the stars in the image were brighter. If a star appeared to be getting brighter, indicating that it might be a lensing event, it was flagged and examined more closely.

In any experiment, one of the hardest challenges is understanding what are called background events—things that masquerade as the events you're looking for, but are caused by something else. An earthquake detector housed underground in San Francisco, for example, needs to filter out vibrations caused by heavy trucks rumbling overhead or reverberations from a nearby building demolition.

In MACHO searches, the main background comes from variable stars—stars that naturally vary in brightness. Variable stars have been observed for hundreds of years, and many different kinds of variable stars are known to exist. In designing MACHO experiments it is essential to develop methods to weed out as many of these stars as possible in order to identify stars whose change in brightness is due to lensing.

Variable stars have several characteristics that make this a manageable undertaking. First, they usually change color (because of a change in temperature) when they vary. Second, the variation is usually not symmetrical in time (i.e., they reach their maximum brightness very quickly, then fade back to their original brightness more slowly). Finally, they often vary more than once.

Another significant background is due to distant supernovae in galaxies that lie hundreds of light-years behind the LMC. These exploding stars, which vary in brightness over the course of a month or so, can be identified by the time evolution of the change in brightness, or by the detection of the galaxy in which the supernova resides.

In contrast, in a straightforward MACHO event the source star should not change color, because lensing affects all wavelengths (colors) of light equally. MACHO events should also be symmetrical in time—the dimple in spacetime caused by the MACHO is symmetrical (a sphere), so as the MACHO passes in front of the star it should get brighter and then dimmer in exactly the same way. Finally, the odds of a given star being lensed by a MACHO in a 1-year period are one in a million, so the odds of the same star's being lensed twice are essentially zero (one in a trillion). Stars that appear to undergo two or more lensing events can safely be assumed to be some kind of variable star.

The MACHO experiments were thus designed to take images using two different color filters—red and blue—on the telescopes. Each star is imaged in these two colors, and the observers create a "light curve" in both colors for every star, making a plot of how the light from the star changes over time (see Illustration 5.5b). In order to be flagged as a possible MACHO event, a light curve must be perfectly flat except for one bump—one time when the star increases in brightness and then decreases back to its original intensity. Any star that shows more than one bump in its light curve is rejected as a lensing event. The bump also has to be exactly the same in both colors. A star that changes brightness more when viewed with the red filter than with the blue is probably a variable star and not a MACHO event, and is also rejected. Finally, events that pass these two cuts are looked at more carefully. The light curve is examined in detail to see if the bump is symmetrical in time, or if any other features would make it unlikely to be a lensing event.

Of course, nature is not quite as kind (or as simple) as described here. Over the course of the MACHO experiments many modifications were made. A new kind of variable star was discovered that initially passed the cuts for a MACHO event. These aptly named "bumper stars" do not increase their brightness by much and change color only slightly, making them hard to pick out from the desired lensing events. On the other hand, some likely lensing events were initially excluded because they appeared to change color during the magnification. This can happen when the source star is actually two stars that are too close together to be seen as separate stars from Earth but only one of these stars is lensed.[8] If the lensed star is much redder than the unlensed star, for example, the combined light of the two stars will be redder at the peak of the lensing event (when the light from the red star has been magnified) than before and after.

In spite of these challenges and against what seemed to be overwhelming odds, success came quickly. In September of 1993, the MACHO and EROS teams announced the first observations of gravitational microlensing,[9] with MACHO reporting one event and the EROS team reporting two possible detections. The first MACHOs had been "seen."

DETECTION

The MACHO experiment had been staring at 1.8 million stars for 1 year and the EROS team had been watching 3 million stars for 3 years when they first presented their results announcing the detection of MACHOs.

Given the incredible experimental challenges and the potentially far-reaching implications of the data—the detection of Galactic dark matter—these results, and those that followed, were carefully scrutinized by the scientific community. Were these events really due to lensing by MACHOs? What did this mean for the composition of the dark matter in the Galaxy? Were WIMPs ruled out as a dark matter candidate?

The light curves that were first presented were of varying quality. Poor visibility from the Mount Stromlo Observatory on some nights—in addition to nights when the Moon was full, making the sky too bright to take images of the LMC—meant that the MACHO data were missing in key sections. The EROS team was also hampered by the fact that it was using photographic plates instead of digitized images at this point.

Nevertheless, the results looked like lensing events. The light curve for the first MACHO event was striking. During the early months of 1992, a star in the LMC increased in brightness until it was seven times brighter than it was before (and after) being lensed (see Color Illustration 3). The event lasted a total of 34 days, from the time the star first appeared 30% brighter than usual to the time it fell back below this same value. This time frame provides an important clue as to *what* the MACHOs are.

The length of time that the star's light appears brighter is the time that it takes the lens to pass over the star. In our magnifying-glass analogy, it is the amount of time that the lens is in front of the star. This time depends on how fast the lens is moving and how big it is. A bigger lens will be in front of the star for a longer time than a smaller lens moving at the same speed. The size of the lens (the Einstein radius) is determined by the mass of the MACHO, which is crucial to determining what the MACHO actually is. Very light MACHOs may be planets; very heavy MACHOs are more likely to be neutron stars or black holes.

To find out, then, what MACHOs are, we need to know the size of the lens. In principle, the size can be calculated from the time duration of the event if we know the speed and exact location of the MACHO. The time duration is measured by the experiments for each event, but in general we know only the average speed with which MACHOs move through the halo (this is determined by the overall mass of the Galaxy), not the exact speed of any given MACHO. Furthermore, we usually don't know exactly where the MACHO is (its distance away from us) and can only estimate the probability of finding a MACHO at a given distance.[10] Thus, we can determine only an average lens size, and a corresponding average mass for the MACHOs based on many lensing events. The more events we have, the more accurately we can estimate the average MACHO mass.

The mass corresponding to an event that lasts for 34 days is about one-tenth the mass of the Sun, assuming that the MACHO is moving at the average speed. Factoring in the uncertainties of the experiment with the uncertainty of the actual speed of the MACHO allowed the MACHO team to place the mass of its first MACHO detections in the rather broad range of between 0.5% and 20% of the Sun's mass. More data would help to narrow down this range.

All three experiments continued to take data, and the evidence that they were truly seeing the invisible—MACHOs—continued to mount. By the end of the decade the combined tally of lenses was about 20 microlensing events from observations of stars in the LMC and SMC.

WIMPS RULE

The MACHO experiment ended on schedule in 1999, when the last observations were made. After monitoring 12 million stars for almost 6 years, the final count from the MACHO team was 13 lensing events toward the LMC. Members of the EROS collaboration watched 33 million stars for 6½ years, but they limited their analysis to a subsample of 7 million bright stars—and found no lensing events. They observed one lensing event of a star in the SMC, which was also seen by the MACHO team.

The final paper of the MACHO collaboration, published in 2000, concluded that a Galactic halo consisting entirely of MACHOs was now ruled out, and estimated that about 20% of the Galactic halo was in the form of MACHOs. The EROS team preferred to present its results as an upper limit on the number of MACHOs in the halo, with no more than about 8% of the halo in MACHOs having masses of about one-tenth to one times the mass of the Sun. A combined analysis of the two experiments showed that, within the uncertainties of each experiment, they are consistent with each other and that less than 20% of the halo is in the form of MACHOs.

MACHOs, the least exotic candidates for dark matter, have now been effectively ruled out as the main component of the dark matter, leaving WIMPs to dominate the Galaxy.

Nevertheless, there seems to be evidence for some MACHOs in the Galactic halo, even if not enough to be interesting from a dark matter point of view. What are these MACHOs—and do we care anymore?

The clues to the nature of MACHOs come from estimates of their mass. On the basis of the length of time that the lensed stars appeared brighter, the MACHO team found that the average mass of a MACHO is about one-half the mass of the Sun. The most obvious candidate for MACHOs with this mass are white dwarf stars, the burned-out cores of stars like the Sun. Unfortunately, white dwarfs are heavy polluters. In the process of becoming a white dwarf, about three-fourths of the original star is flung out into the surrounding area. This material contains large amounts of the by-products (elements such as carbon and oxygen) that are produced when a star burns its hydrogen fuel. If 20% of the halo is in the form of white dwarfs, we should be able to detect huge quantities of these elements in the gas clouds in our Galaxy (and others). Although we do see these elements, we don't find nearly enough to be consistent with so many white dwarfs being formed at an earlier epoch in our Galaxy's history. At most, less than 5% of the halo can be in the form of old white dwarfs.

Many ideas have been suggested to explain the MACHO events, including fewer white dwarfs in a small shroud surrounding the disk of the Galaxy; lenses that exist in the LMC itself instead of our Galaxy; or black holes that formed in the very early Universe.[11] All of these ideas have problems, and the answer simply isn't known yet. But the answer may yet be important for our understanding of other questions at the forefront of cosmology, especially if MACHOs turn out to be white dwarfs or black holes. Either case would be the equivalent of finding fossils of a much earlier period in the Universe, allowing us to test theories of galaxy formation or the creation of protons from quarks. We hold in our hands a shard, a fragment of a bone, hoping that it's part of a new kind of prehistoric creature—a missing link in our understanding of the cosmos. But until we have more data from future MACHO searches, we're not yet completely convinced it's not just an oddly shaped piece of stone.

CHAPTER 6

Black Holes and Planets

WIMPs may constitute the bulk of the dark mass in the Galaxy, but the census of dark inhabitants is far from complete. Hidden among the forest of stars that comprise the visible Galaxy are dark objects that are as intriguing as they are elusive: black holes and planets. Both are notoriously difficult to detect—and both have recently been found by the newest member of the search team: gravitational lensing.

Gravitational microlensing experiments have redirected their telescopes toward the center of the Milky Way, peering through the disk of stars and gas clouds to search for the small warps in spacetime induced by black holes and planets. It has proven to be fertile ground. In January 2006 the prestigious journal *Nature* announced the discovery of a new extrasolar planet—with a mass roughly five times the mass of the Earth, it was at that time the smallest planet known to exist outside of our Solar System. Separated from its home star by roughly three times the distance between the Earth and the Sun, this sub-Neptune-sized planet and its rather typical star lie halfway across the Galaxy, 21,000 light-years away. It was found via gravitational lensing, the only technique currently able to spot such a small planet in an Earth-like orbit about a normal star.

Microlensing searches also broke through another barrier in the search for black holes. The disk of the Galaxy should be home to millions of stellar-mass black holes, the corpses of stars much larger than the Sun. The majority of stars in a galaxy are formed, live, and die within the disk, and their remains are likely to be found there as well. But black holes are invisible—their location is usually betrayed by the effects they have on a stellar partner that can be seen. Lone black holes, roaming the Galaxy unaccompanied by a visible stellar buddy, are impossible to find. Or they were. In 1999, a black hole, flying solo, triggered the longest microlensing event on record.

BLACK HOLES

To a theorist, a black hole is a black hole. It can be characterized by just three features: its mass, its spin, and its charge. Astronomers take a different view of black holes, lumping them into two main categories based on their size and origin: *stellar black holes* and *supermassive black holes*. The most numerous black holes are likely to be a few or a few tens of times as massive as the Sun—remnants of an earlier population of stars. These stellar black holes are much harder to detect than their behemoth cousins, the supermassive black holes that dwell at the centers of most galaxies, including our own. Supermassive black holes are millions or even billions of times the mass of the Sun, and they are thought to be the engines that power quasars in the early Universe.

A black hole of any size is formed whenever too much mass is compacted into a very small space. The definition of *very small* depends on the amount of mass you have to work with. To create a black hole with the mass of the Earth, you simply need to cram the entire Earth into a ball about two-thirds of an inch across. Or you could start with the Sun and squeeze all of its mass into a sphere just over 3½ miles in diameter.

In spite of this straightforward recipe, no one has yet manufactured a black hole in the laboratory, and it seems unlikely that anyone will in the foreseeable future, but we have a pretty good idea of how nature manages the technical details. The mass compactor of the Universe is, of course, gravity. Given enough mass, gravity will cause matter to collapse in on itself—until and unless some other force is strong enough to stop it. A star continues to shine as long as the gas pressure generated by the heat of nuclear reactions opposes the gravitational infall of the star's mass. When the star runs out of its nuclear fuel, gravity takes over again, further compressing the material in the dead star until it becomes so dense that the electrons within it start to object. Electrons strongly resist being squashed too closely together—courtesy of the Pauli exclusion principle, a fundamental tenet of quantum mechanics asserting that no two electrons can occupy the same state. This electron pressure can hold out against further collapse if the mass of the dead star is less than about one and a half solar masses, producing a state of gravitational détente known as a *white dwarf*.

Stellar remnants with more than the white dwarf mass limit will override the objections of the electrons. Their self-gravity continues to squeeze the mass together until the increase in pressure forces each electron to combine with a proton, converting almost the entire mass into neutrons.[1] These neutrons—which also aren't too thrilled about sharing such close quarters—present a united front to oppose further collapse. If the mass of the ever more compact stellar remnant is less than about two or three solar masses, the contraction stops here and a neutron star is formed.

Even neutrons can't hold out against the most massive stellar remains. If the mass of the dead star exceeds three times the mass of the Sun, the efforts of the neutron pressure are futile. At such masses gravity is unstoppable. It will continue to pull the mass inward on itself past the point of no return, creating an object of infinite density, a warp in spacetime so severe that not even light can escape from within it—a black hole.

Dark Stars

The idea of an object so massive that light would be unable to escape from its gravitational clutches was actually first proposed long before Einstein started fiddling with light beams and gravity. In 1783, a British scientist, the Reverend John Michell, used Newtonian gravity to calculate the escape velocity of a particle of light from the surface of a star.[2] His calculations involved essentially the same equations that determine how fast a rocket must be moving in order to escape from the gravitational pull of the Earth or the Moon. Michell assumed that

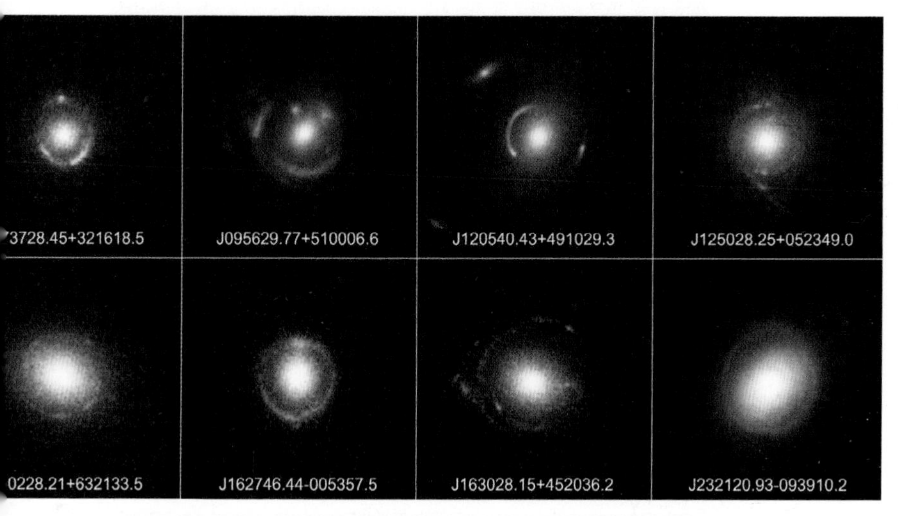

COLOR ILLUSTRATION 1 Einstein Rings. Astronomers utilized the combined power of the Hubble Space Telescope and the Sloan Digital Sky Survey to find eight new Einstein Rings in 2005. At the center of each image is a bright orange-yellow galaxy located a few billion light-years from Earth. Directly behind each of these galaxies is a second galaxy, roughly twice as far away from us, whose light has been bent into the blue ring that appears to encircle the closer galaxy. If the lensing galaxy were not there, the more distant galaxy would appear as a small blue smudge.

COLOR ILLUSTRATION 2 Giant Arc. At the center of this image from the Hubble Space Telescope is a bright yellow-orange galaxy about 6 billion light-years away from us. Just above the center of this galaxy is a much more distant galaxy (roughly 10 billion light-years away) whose light has been stretched and curved into a large blue arc by gravitational lensing, revealing the warp in space induced by the mass of the closer galaxy.

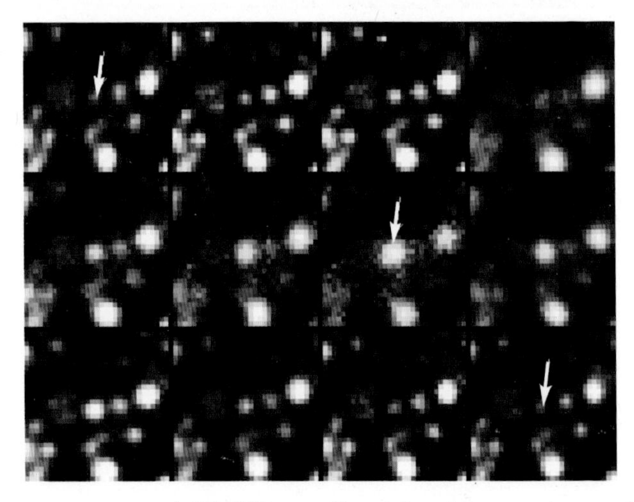

COLOR ILLUSTRATION 3 A MACHO event. From left to right, top to bottom, these 12 panels are sequential images of a field of stars in our closest galactic neighbor, the Large Magellanic Cloud, taken over a period of 90 days. The arrow points to a star that is being lensed by an invisible MACHO. Over the course of about 34 days the light of this star was magnified by the MACHO to a maximum level of about seven times its original value (seventh panel). The first and last panels show the star as it appears before and after the MACHO passes by (note that the star is one of the fainter stars in this field).

COLOR ILLUSTRATION 4. Mapping dark matter. (a) Gravitational lensing by a cluster. The bright yellow-orange objects in this image are galaxies that are members of a cluster of galaxies (CL 0024+1654). The odd-looking blue objects seen at about four, eight, nine, and ten o'clock (and also near the very center of the image) are multiple images of a single distant galaxy that is not part of the cluster, but billions of light years behind it. The mass of the cluster acts as a lens to bend and distort the light from this distant galaxy—stretching, magnifying, and multiplying the original to create five copies. (*See Illustration 7.1 on page 147 for annotated version.*)

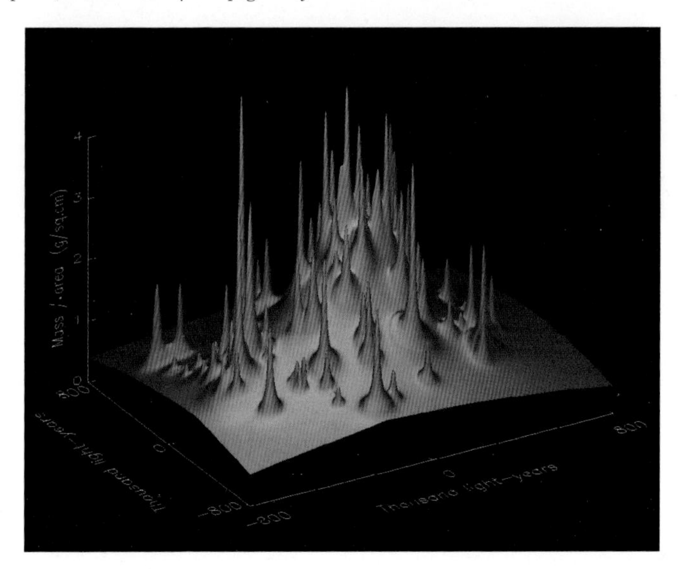

(b) A complex lens. Although only the galaxies in CL 0024+1654 are visible in the Hubble image above, most of the mass in the cluster is in the form of invisible dark matter. Shown here is a map of the total mass in the cluster, constructed from the observations of multiple images of a single blue galaxy created by gravitational lensing. The sharp peaks correspond to the mass and position of individual galaxies in the cluster; the underlying mountain represents the dark matter that comprises the bulk of the cluster mass.

COLOR ILLUSTRATION 5 Quasar quintuplets. The bright orange galaxy at the center of this image taken by the Hubble Space Telescope is the central galaxy of a cluster (known officially as SDSS J1004+4112) lens that has created five images of a single distant quasar (four can be seen as the bright, white, star-like objects at four, eight, eleven, and twelve o'clock about the central galaxy; the fifth is hidden in the glare of this galaxy) and multiple images of several other distant galaxies. (*See Illustration 7.2 on page 150 for details.*)

COLOR ILLUSTRATION 6 A massive lens. Faint blue and red arcs can be seen circling about the center of cluster Abell 1689, one of the largest and most powerful gravitational lenses found to date, creating an almost Van Gogh–like effect. Over 100 arcs have been identified—multiple copies of 30 distant galaxies. The bright yellow objects are galaxies in the cluster, which is 2.2 billion light-years from Earth; the bright white spiked objects are stars in our own Galaxy.

COLOR ILLUSTRATION 7 A cosmic magnifying glass. The mass in cluster Abell 2218 creates an enormous dent in spacetime, which acts as a gravitational lens. This lens stretches and magnifies the images of galaxies that lie behind it, affording scientists a view of some of the most distant galaxies ever seen, over 12 billion light-years away. Without the magnifying power of the cluster lens, these ancient galaxies would be too faint to be detected even by the Hubble Space Telescope.

COLOR ILLUSTRATION 8 The Bullet cluster. This image is a composite created from a Hubble Space Telescope image (optical light), X-ray observations of the same region of the sky (pink), and gravitational lensing data (blue). The Bullet cluster is actually two clusters that have recently collided with one another, and are visible in optical light as a large group of galaxies just to the left of the center of the image, and a smaller group to the right. (Bright spiked objects are stars in our own Galaxy.) Overlaid on this image is the distribution of hot gas detected by X-ray telescopes and shown here in pink. In blue is the distribution of matter in the clusters derived from gravitational lensing studies.

COLOR ILLUSTRATION 9 A computer simulation of the cosmic web. A slice of the simulated universe roughly 68 million light-years thick and 1.7 billion light-years across. At the center is a bright spot representing a cluster of galaxies. (a) Normal matter only (galaxies, clusters, and gas).

(b) Dark and normal matter. The intensity in this image corresponds to density of matter, not light—most of the matter is dark matter, and thus invisible to our telescopes.

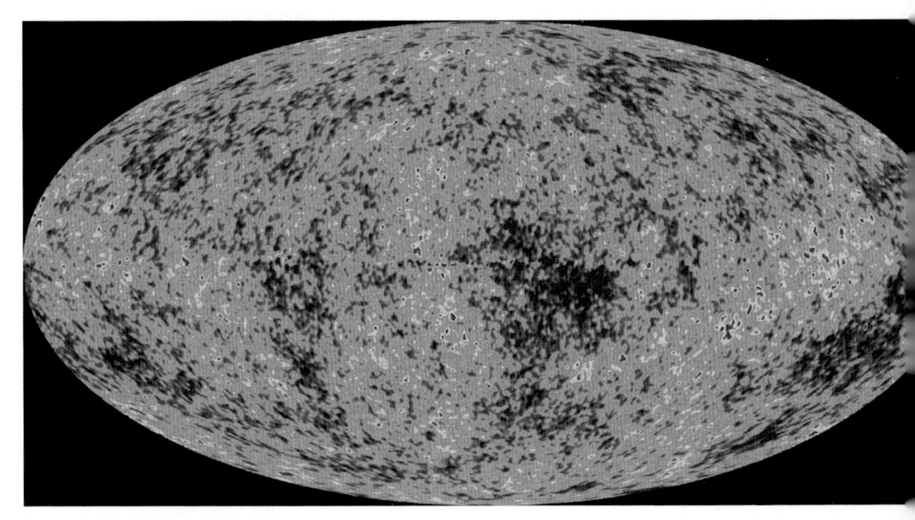

COLOR ILLUSTRATION 10 The afterglow of the Big Bang. An all-sky map of the light from the Big Bang—the cosmic microwave background—presents us with a picture of the Universe as it appeared over 13 billion years ago, when the Universe was only 380,000 years old. The colors represent tiny variations in the temperature of this light (red spots are hotter; blue are colder), with a range of ± 200 millionths of a degree Kelvin.

light is composed of massless particles—a theory that was popular at the time—and found the parameters of a "dark star" for which the escape velocity is greater than the measured speed of light. Light would never be able to outrun the gravitational influence of such a star, just as a ball or rocket moving upward with less than the escape velocity will always return to Earth. What goes up (with less than the escape velocity) must come down.

Then along came Thomas Young. As the eighteenth century gave way to the nineteenth, Young performed a series of experiments establishing that light behaves as a wave. These results dampened enthusiasm for the particle theory of light (at least temporarily—current quantum theory says that light is both particle and wave), and dark stars disappeared from the scientific lexicon for the next 100 years or so.

They snuck back in under cover of the First World War. In November of 1915, a middle-aged German soldier stationed on the Russian front interrupted his calculations of ballistic trajectories to read the latest issue of the *Proceedings of the Prussian Academy of Sciences*. Karl Schwarzschild, a well-known astrophysicist (and over the age of 40), volunteered for military service when World War I broke out but continued to squeeze in as much science as he could. Just days after reading Einstein's new theoretical description of space, time, and gravity, Schwarzschild produced the first exact solution to Einstein's equations, calculating the curvature of spacetime produced by a single spherical, massive body.[3]

Recognizing that his military buddies might not be the most receptive audience for a paper on general relativity, Schwarzschild mailed the results of his calculations to Einstein. Einstein was very impressed with Schwarzschild's work, and having had the good sense not to volunteer for the kaiser's army, was free to present the solution to the Prussian Academy of Sciences on Schwarzschild's behalf.

Contained within Schwarzschild's missive from the Russian front was the blueprint for a black hole. Schwarzschild's solution to Einstein's equations applies to relatively plebeian astronomical objects like stars and planets, but it also allows for the existence of a massive body from which even light cannot escape. According to his calculations, for a given mass there exists a critical radius. An object whose mass is compacted within this "Schwarzschild radius" will create a deformation in spacetime so extreme that nothing can escape from within it. Neither Einstein nor Schwarzschild considered such an object—which they labeled with the catchy moniker *Schwarzschild singularity*—to be a physical possibility. They quickly dismissed this singularity (the term refers to a point where a theory produces nonsense—the equivalent of dividing the number 5 by 0, for example) as a curio of theoretical interest only. Surely nature had ways to prevent such an implausible confection.

Unfortunately, Schwarzschild became ill soon after his correspondence with Einstein and was invalided out of the army just a few months later. He died in May of 1916, over 50 years before his singularity was rechristened by John Archibald Wheeler[4] with a much more palatable name—as of 1967 Schwarzschild's singularities were henceforward known as *black holes*.

Extreme Sports Edition

Black holes are strange beasts. Scientists are nowhere near completing their theoretical dissection of these bottomless pits in spacetime, but just what do we know about them? Far from the event horizon, a black hole is gravitationally equivalent to a star of the same mass. To experience the truly outlandish behavior of black holes, you need to get up close and personal.

Those who enjoy extreme sports might consider black hole bungee jumping—a unique treat for both participant and fans. Just make sure the bungee cord keeps the jumper above the event horizon—the imaginary sphere around the black hole defined by the Schwarzschild radius—and don't forget to use the full relativistic theory when calculating the length of the cord. The brave soul at the end of the bungee cord will appear to someone watching from the sidelines to become ever redder and fainter as he falls closer and closer to the event horizon, his cries for help dropping in pitch like an audiotape played at an ever slower speed. If the bungee cord is even a millimeter too long and the now-doomed athlete passes through the event horizon, he will eventually fade from view completely. He won't be bouncing back.

From the point of view of the jumper, the ride into the black hole is fascinating, at least in the beginning. As the event horizon approaches, strange things begin to happen to his view of the Universe. The gravitational lensing properties of the black hole focus the light from the entire sky—all of the visible stars and galaxies—into an ever smaller disk over his head (the direction opposite the black hole), eventually shrinking to a single dot of light while the rest of the sky becomes completely black. He won't notice anything special as he moves through the event horizon, but at some point (the exact location depends on how massive the black hole is) the trip will become more than a little uncomfortable. The unpleasantness comes from the fact that the pull of gravity is not constant but depends on where he is, getting stronger the farther into the black hole he falls, just as the strength of gravity is stronger at the surface of the Earth than it is 10 miles up. Inside a black hole this difference in strength is much larger over a much shorter difference. If he is falling in feet first, the pull on his feet is stronger than the pull on his head, resulting in a rather painful process of elongation—he will be stretched into a long thin strand of human spaghetti. This is a mercifully brief state of affairs, since he will soon thereafter be shredded apart completely, putting an end to the gravitational fun and games.

A black hole that is spinning makes for an even wilder ride, as the black hole drags spacetime (and anything traveling in that spacetime) around with it. Think of it as the ultimate swirly. The end of the trip is always the same, however. Anything and everything that falls into a black hole is lost forever to the Universe in which we live.

This is not to say that a black hole never loses any weight. Stephen Hawking discovered that black holes are not really black. They radiate. Energy in the form of particles of all kinds, including light, streams away from a black hole at a rate that is inversely proportional to the mass of the black hole. Which sounds ridiculous—if light can't escape from a black hole, how can a black hole radiate? Enter quantum mechanics, which states that empty space is never really empty (an idea that will also be important when we discuss dark energy again), but a bubbling cauldron of particles that flicker into existence, and back out again, in the briefest moment of time. These virtual particles are always created in pairs—a particle and its antimatter twin—and they owe their existence to playing the cosmic float. To fund their self-production they borrow energy from the Universe, then pay it back again before the Universe has time to notice. When this pair creation occurs just outside the event horizon of a black hole, the particle twins can be separated, divided forever by the black hole. As one particle zooms away from the black hole, its partner strays across the point of no return and disappears forever beyond the event horizon into the black hole. The escaping particle carries away positive energy in the form of mass and motion, while its doomed partner brings with it an energy deficit, decreasing the mass of the black hole.

The smaller the mass of a black hole, the more intensely it radiates, becoming smaller and less massive in a runaway process whose end point is still up for debate. The black hole might completely disappear in a blaze of radiation, or it might leave behind a tiny remnant, a small scar on spacetime. (The answer—again—ultimately depends on understanding quantum gravity.) Whatever its final state, the last few moments of a black hole should be viewed from a safe distance. A very small black hole, containing the mass of this book, for example, would pose a serious radiation hazard, frying everything in its immediate vicinity. The end-stage burst of radiation of such teensy black holes might even allow us to detect them, although none have yet been seen. However, the stellar-mass black holes we expect to populate the Galaxy are far too large to produce any noticeable Hawking radiation. Black holes with the mass of the Sun evaporate so slowly (the evaporation time for such a black hole is 10^{58} times longer than the age of the Universe) that, for all practical purposes, they are truly black.

The Telltale Star

Because stellar black holes are just that—black—we cannot hope to see them directly through any of our telescopes (except of course, Einstein's Telescope, but we'll get to that later). We need some way to infer the presence of a black hole via its impact on something that we can see. There are two possible effects we can search for: the gravitational tug of a black hole on a nearby object, and the high-energy shriek sent out by something that's been pulled too close.

Before going further, we need to get a certain myth out of the way. Black holes are not cosmic vacuum cleaners. It is certainly true that everything that falls within the Schwarzschild radius will be claimed by a black hole, never to be seen again. But at a respectable distance, the gravitational field of a black hole is no different from a less exotic object with the same mass. Nothing is summarily sucked into a black hole, any more than it would be sucked into the Sun. Just as the planets happily orbit about the Sun, or two stars orbit about each other, a star can orbit about a black hole.

The key difference is that we can't see the black hole. Consider a system of two stars in orbit about one another. By observing their motions we can deduce their masses. If we replace one of the stars with a black hole, the orbital dance will continue, but the movements will seem unbalanced. A single star will appear to be in orbit about nothing.

This observation alone does not provide us with a viable search mechanism for finding black holes. There are billions of stars in the Galaxy, and it's not possible to cross-examine each one for the existence of a black hole partner.

Luckily (for us), black holes are not good company. They're often in the process of consuming any nearby stellar companion, peeling off the outer layers of the doomed star into a whirlpool of gas that spirals down into the black hole. As it spins faster and faster on its final descent, this gas is heated to millions of degrees, emitting a blaze of X-rays before disappearing forever over the event horizon. By searching for bright sources of X-rays, astronomers can home in on locations within our Galaxy that are likely to harbor a black hole.

Thus, the standard operating procedure for finding a stellar-mass black hole is essentially the same method used to detect the first black hole candidate. Astronomers sift through a catalog of X-ray sources in the Galaxy, looking for those with a visible star at the same location.

The motion of this star is then monitored closely for evidence of a dark companion, whose mass is determined from the orbit of the visible star. The identification of the unseen object as a black hole is based on this mass. If the mass of the dark companion is more than three times the mass of the Sun, it is likely to be a black hole.[5]

In 1990 Stephen Hawking paid off a bet he had made 16 years earlier, shortly after the discovery of Cygnus X-1. Cygnus X-1 was the first X-ray source detected in the constellation Cygnus the Swan. It's also the name of the first black hole ever discovered.

Hawking had wagered that scientists would ultimately disprove the existence of a black hole in Cygnus X-1; Kip Thorne, one of the world's leading black hole experts, had placed his chips on the other square, betting on the eventual confirmation of a resident black hole. Although at that point in time Thorne claimed he was only 95% sure of the black hole's existence, he nonetheless graciously accepted his winnings (a magazine subscription), along with a signed concession, from Hawking.

Cygnus X-1 now heads a growing list of about 20 stellar-mass black holes (and another two dozen or so possible black hole candidates that have not yet been confirmed), all of them detected courtesy of their gas-donating stellar companions. Given a detection rate of about three per decade, it's obvious that finding black holes is not easy. Tracking down those without a stellar partner would seem impossible.

Lonely Black Hole

In his 1994 book on black holes, Thorne presents several possible methods for discovering black holes. Before going into the details of the standard technique outlined in the preceding discussion, he makes a brief mention of gravitational lensing, quickly dismissing it as impractical: "The necessary Earth-hole-star lineup would be an exceedingly rare event, so rare that to search for one would be hopeless."[6] Sound familiar? This bet is one that he lost.

The very center of the Galaxy, known as the *bulge*, is home to the densest population of stars. Packed into a barlike configuration over 10,000 light-years long and roughly half as wide, these stars offer an enormous background of light sources for lensing studies. Further, the

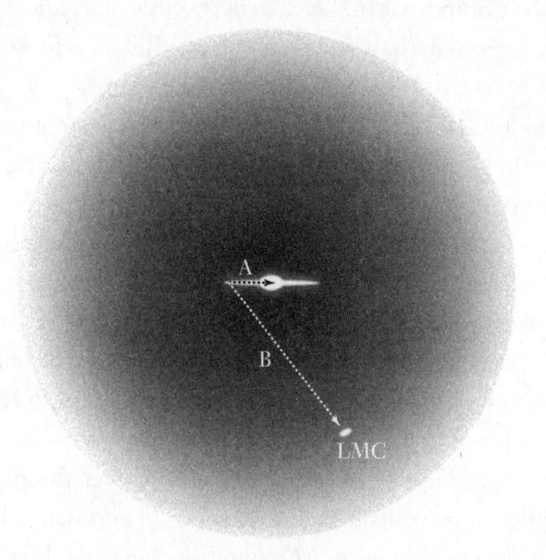

ILLUSTRATION 6.1 Lensing in our Galaxy. MACHO experiments look at stars in a nearby galaxy known as the Large Magellanic Cloud (LMC). These experiments are looking through the dark halo of our Galaxy as shown by the dashed line labeled B. Searches for black hole and planet lenses look through the disk toward the center of the Galaxy, indicated by the dashed line labeled A.

line of sight to the stars in the bulge pierces through the huge numbers of stars in the disk of the Galaxy and the near side of the bar, all of which are potential lenses. It's a winning combination for microlensing searches—lots of lenses moving past lots of light sources guarantee that the odds of observing a lensing event go way up.

Three major collaborations—MACHO, OGLE (Optical Gravitational Lensing Experiment), and MOA (Microlensing Observations in Astrophysics)—are taking advantage of this lensing bonanza and logging a total of about 500 microlensing events each year. Most of the lensing events these groups record are due to the lensing of one star by another—a star in the disk moves almost directly in front of a more distant star in the bulge, for example—but if significant numbers of black holes are whirling through the disk of the Galaxy along with the stars, they, too, will act as lenses. As Thorne pointed out, the requisite collinear positioning of the black hole, source star, and Earth-based observer is rare, but (like Einstein) he was bit overly pessimistic in declaring it hopeless. In fact it's already been seen.

Both the OGLE and MACHO groups were watching as a star in the Galactic bulge was slowly magnified to 32 times its original brightness, then just as slowly returned to normal. The entire event lasted over 3 years—the longest microlensing event ever observed. The final word on the exact mass of the lens is not yet in (estimates range from about 4 to over 100 times the mass of the Sun) but the several different analyses of the data all arrive at the same conclusion: the lens responsible for this event, known as MACHO-99-BLG-22/OGLE-1999-BUL-32, is a black hole.[7]

It is the only lone stellar-mass black hole ever discovered. (Two other microlensing events have been proposed as possible black holes, but the data are not conclusive.) Gravitational lensing does not require that a black hole have a stellar partner in order to be detected—the warp in spacetime induced by the black hole itself gives it away.

The microlensing teams look for two key signatures of a black hole: the lens must be massive and it must be dark. As with other black hole searches, the mass of a candidate must be well above three solar masses in order to convince a reasonably skeptical scientific audience that a black hole (and not a neutron star) has been spotted. Once the search has homed in on the heaviest lenses, there are only two possibilities—a black hole or a very massive star. Stars with masses more than three times the mass of the Sun should be relatively easy to spot even halfway across the Galaxy. Follow-up observations of the lens (or the region of the sky where the lens should be) can either detect or eliminate the possibility of a massive stellar lens. If you can see it, it's not a black hole.

Determining the mass of the lens is the hard part. For the simplest microlensing events, all of the information about the lens—its mass, its location, and its speed—is encoded in one bit of data: the length of the event. A measurement of the event duration tells us only about a combination of the mass, speed, and location of the lens, and it's impossible to separate out any one of these numbers, including the mass, without more information.

In searching for halo MACHOs we were content to play a statistics game with the mass. We have a rough idea of where the MACHOs are—the general distribution of MACHOs in the Galactic halo is mapped out by the motions of the stars, as shown in Illustration 5.3. These motions also give us the total mass in the Galaxy, which in turn tells us how fast, on average, we expect the MACHOs to be moving. We can thus estimate the most likely speed and location for the lenses. By averaging the data for many halo lensing events, we can determine an average mass for MACHOs in the halo, which is enough to clue us in to what the MACHOs might be.

For black hole detection, however, averages are no longer sufficient—we need to know the mass of an individual lens in order to classify it as a black hole. This requires searching for microlensing events that are more complicated than the simplest cases discussed so far; events that provide more information about the lens than just the event duration.

The Blink of an Eye

One of the most basic techniques in astronomy is presented in every high school astronomy course, in part because it's so easy to demonstrate. Hold up a finger a few inches in front of your nose. Close one eye; then open it and close the other. Blink back and forth a few times and watch as your finger appears to jump from one spot to another. Now hold your finger an arm's length away and repeat the exercise. Your finger will still appear to hop back and forth, but its excursions will be smaller. This is *parallax*. Because your eyes are separated by a few inches, your point of view changes depending on which eye you use. Your finger isn't moving, but its position relative to distant objects (the picture on the opposite wall of the room for example) changes, making it seem as though the finger is hopping from side to side. The closer your finger is to the front of your face, the more pronounced the effect is. Parallax—the apparent change in position—can be used to measure distance.

In order to detect parallax effects for objects at astronomical distances away from us, we need to observe them from two different viewpoints that are also separated by astronomical scales. Until recently, we were pretty much stuck here on planet Earth—even the

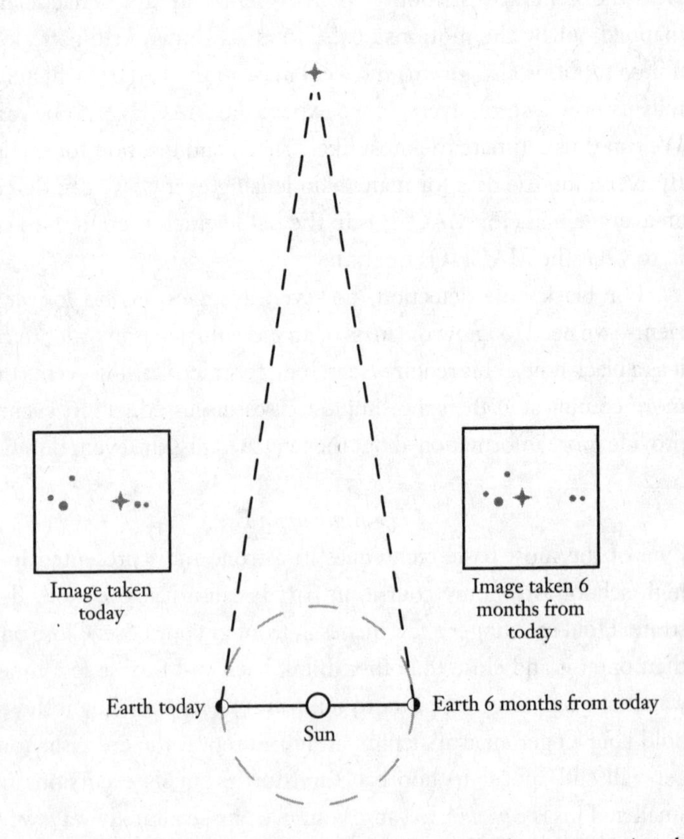

ILLUSTRATION 6.2 Parallax. The position of a relatively nearby object, seen against the backdrop of much more distant stars, shifts as the Earth orbits about the Sun.

Hubble Space Telescope is only about 350 miles above the Earth's surface. (For comparison, the Moon, our closest astronomical companion, orbits at a healthy distance of almost a quarter of a million—238,900—miles.) But parallax has been a staple in the astronomical tool kit since 1838. Galileo even attempted to use it over 200 years earlier, failing only because of the limitations of his telescope. He understood the basic principle very well—because the Earth is in orbit about the Sun, the Earth itself changes location. The Earth is one astronomical unit—about 93 million miles—from the Sun. Thus, the position of the Earth in June is roughly 186 million miles away from its position in January, when it has traveled to the opposite side of the Sun—a large enough separation to detect parallax for an object hundreds (or, as we'll soon see, even thousands) of light-years away.

What does this have to do with microlensing experiments? In the simplest lensing scenario, the motion of the lens in front of a star induces a time-varying change in the apparent brightness of the star. The motion of the Earth adds a new complication to this process. When the Earth moves, our line of sight to the star pierces through a different part of the lens, giving us a slightly different view through the gravitational lens in June than we have in January. This shift in viewpoint will also change the apparent brightness of the star, since the magnifying power of the lens depends on which part of it we're looking through. It will usually be a small effect, especially in comparison with the change in brightness due to the motion of the lens in front of the star, which is responsible for most of the observed increase.

The source star is also moving, of course, but in general its motion can be ignored because it's so far away. The farther away an object is, the harder it is for us to detect its motion. An airplane flying overhead at 25,000 feet seems much slower than the same jet as it buzzes by at 2,000 feet. If the source star's motion does become important, its impact on the light curve is known as *xallarap*. Detailed lensing calculations take into account the motions of all three—Earth, lens, and star.

The motion of the Earth induces a departure from the shape of the standard light curve that was discussed in Chapter 5, where the MACHO moved with a constant speed across our line of sight to the source star. In effect, the Earth's motion leaves an imprint on the light curve, marking it with a known velocity, much as the key on a map indicates the scale of the map (e.g., 1 inch = 10 miles). This allows us to extract from the data the relative velocity of the lens (how fast it's moving relative to the star being lensed), which in turn narrows down our estimate for the mass of the lens.

For this parallax effect to be detected, the lensing event must last at

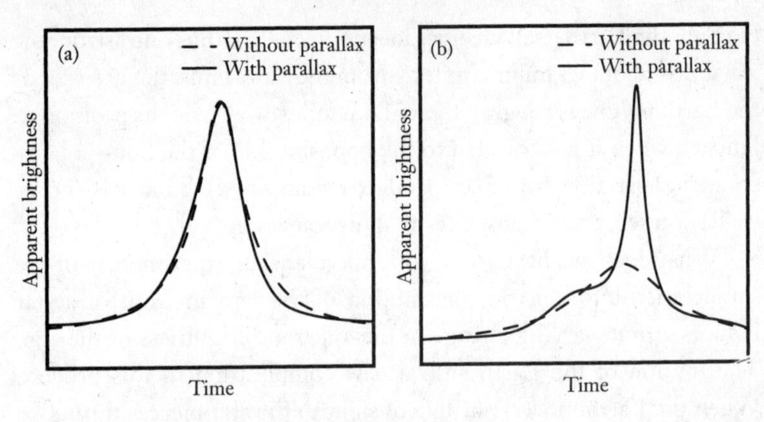

ILLUSTRATION 6.3 The effect of parallax. The apparent brightness of a star changes over time as a lens passes in front of it—the exact time course of the change depends on the details of the lensing event. The dashed line in both panels represents the simplest case of a lens moving in front of distant star (also shown in illustration 5.5 on page 102). However, if the lens is relatively close to us and the change in brightness takes place over a long time period (on the order of a year or more), the light curve will deviate from the simple case because of the effect of parallax—our point of view through the lens changes as the Earth moves around the Sun. Two examples illustrating the effects of parallax are shown by the solid line in each panel. (a) The shift in a light curve caused by parallax is usually fairly subtle; (b) however, some observed parallax lensing events depart dramatically from simple lensing, resulting in a more complicated light curve. Such striking effects can be caused by a very long event in which the lens lies almost directly over the star at some point in time.

least several months, giving the Earth time to move a significant distance. This timing dovetails beautifully with the search for black holes. A black hole in the disk of the Galaxy, weighing in at four or more solar masses, will typically produce an event that lasts a year or more—long enough for the Earth's motion to leave its imprint on the light curve.

As Illustration 6.3a shows, the effects of parallax are often subtle but detectable; occasionally parallax can also induce more exaggerated changes, as in Illustration 6.3b. Astronomers have combed through the thousands of microlensing events collected to date, checking each one for evidence of parallax. A computer program generates a huge number of theoretical light curves, varying the amount of parallax, as well as the mass, distance, and speed of the lens, in each one. It then compares the data to these light curves, and selects those for which parallax provides the best fit.

Once the events that exhibit parallax have been culled from the herd, the light curves are studied in more detail. The detection of any parallax puts an upper limit on the distance between Earth and the lens (the effect of parallax diminishes with distance, so if the lens is too far away no parallax will be seen), while the amount of parallax—how strongly the light curve is distorted—places a lower limit on the distance. (The closer the lens, the stronger the parallax.)

Combining the observed event duration and amount of parallax, astronomers can now extract the mass of the lens as a function of the distance to the lens. The Galaxy itself gives us the final clue we need to narrow down the search. Objects in the disk or the bulge of the Galaxy are orbiting about the Galactic center with velocities that are, in general, dictated by how far they are from the Galactic center—their orbital speed is prescribed by gravity. If this information is folded into the mix for the first black hole lens (MACHO-99-BLG-22/OGLE-1999-BUL-32), the data point to either a relatively nearby (5,000 light-years or so) black hole on the order of about 100 times the mass of the Sun, or a four-solar-mass black hole close to the Galactic center. Either way, it's a black hole.

Without more information than even the detection of parallax can provide, the masses of black holes discovered via gravitational lensing will remain uncertain. The answer to this problem lies in space. The Space Interferometry Mission (SIM) is a proposed NASA mission that is nearing completion of its preliminary design. The SIM satellite will carry two optical telescopes into an orbit that will increase in distance from Earth by about 9 million miles each year. One of its stated goals is to "make a complete census of the stellar population of the Galaxy, including both ordinary stars and dark stars."[8] It intends to find and weigh these dark stars (black holes) using gravitational microlensing.

Because it will reside in a large orbit, SIM can be used in combination with an Earth-based telescope to obtain parallax information about a microlensing event. But its true power lies in its ability to see the lensing event much more clearly than ever before. In general, a larger telescope can produce a clearer image than a smaller one. This increase in clarity can also be achieved by combination of the light from two or more smaller telescopes in a process known as *interferometry*. By comparing the light from its two telescopes, SIM can zero in on the image of a lensed star with unprecedented precision. Even SIM will be unable to resolve this image into the two images produced by the microlensing (recall that, in microlensing, the two images produced by the lensing are seen as a single, brighter image), but it will see something that ground-based telescopes cannot. In microlensing, the center of the combined image is shifted very slightly from the center of the unlensed star as the event proceeds. SIM will be able to detect this tiny shift—the final piece of information needed to determine the speed, the location, and the mass of the lens. Black holes will no longer be able to hide.

PLANETS

If finding black holes is difficult, finding planets is equally challenging. The first extrasolar planets were discovered in 1992 by Aleksander Wolszczan, who was monitoring the radio light beamed our way by a spinning neutron star known as a *pulsar*.[9] A pulsar sends out a beautiful staccato of light pulses in a rhythm set by the rotation of the neutron core. Wolszczan detected little blips in the timing of this rhythm, tiny variations in the otherwise smooth revolution of the pulsar caused by the slight gravitational tugs of planets (two or three in this case) in orbit about it.

These planets are not the most desirable real estate—the radiation given off by the neutron star would fry any form of life foolish enough to attempt a visit—but then neither are most of the other extrasolar planets found to date. Astronomers have searched around stars of a less lethal variety—stars more or less like our Sun—and discovered over 375 planets to date. None appears to be conducive to life. In general, they are too big (more like Jupiter, a gas giant, than the rock we call Earth), and too close to their home star for comfort.

It's not that most planets in the Galaxy are gas giants on tight orbits. Rather it's the old story of looking for lost keys at night under the streetlight simply because that's the only place you might be able to see them. Current planet-finding techniques are most sensitive to planets that are relatively heavy and close to the star about which they orbit, so it should be no surprise that these are the kinds of planets we're finding.

The most prolific method of spotting planets relies on detecting the gravitational influence of the planet on its star. Two (or more) objects locked in orbit about one another are bound by their mutual gravitational attraction. Stars tug on their planets—and planets tug back on their stars, each with a strength proportional to its mass. Because the gravitational muscle of a planet is relatively weak compared to that of the parent star, the planet's impact on the motion of the star is also pretty feeble. Nonetheless, the presence of a planet will cause a star to wobble ever so slightly, moving back and forth to the beat of the planet's orbit—one wobble for each trip of the planet about the star. The larger the planet, or the closer it is to its star, the more pronounced the wobble—and the more easily it can be detected.

Teasing out the tiny oscillations in a star's motion that reveal the presence of a planet is extremely difficult. Motions of distant objects in the Universe are usually measured by *Doppler shifts*. As you wait at a railroad crossing gate with your car window open, the sound of the train whistle shifts in pitch as it passes by—it sounds a higher note (a higher frequency) as the train barrels toward you, then shifts to a lower pitch as the train speeds away. Light can also be Doppler shifted. If the source of the light is moving toward you, the frequency of the light you detect will be higher than if the source is stationary (relative to you); if it's moving away, you'll detect a lower frequency, just as you hear the pitch of the train whistle drop to a lower note. Different frequencies of sound correspond to different notes on the musical scale; different frequencies of light correspond to different "colors" on the electromagnetic spectrum. In the visible part of the spectrum, red is lower frequency, blue is higher.

The oscillations of a planet-bearing star are far too tiny to produce a noticeable change in the overall color of the star, but this motion can produce a subtle shift in the spectrum of the star's light. Just as the light from a standard lightbulb can be spread into a rainbow pattern by a glass prism, the light from a star can be dispersed into a spectrum of colors. Imprinted upon this spectrum are a series of lines that corre-

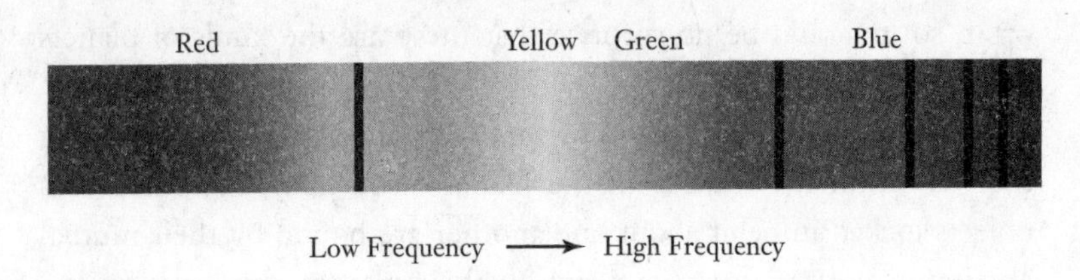

ILLUSTRATION 6.4 Spectrum of light. The dark lines that appear in this spectrum of light in optical wavelengths (from red on the left to blue on the right) are created by the absorption of specific colors of light by hydrogen. This particular pattern of lines is analogous to a fingerprint for hydrogen—other elements each produce their own characteristic set of lines.

spond to various elements in the star's atmosphere (see Illustration 6.4). Hydrogen has one set of lines; calcium has a different set; iron yet another. Like a fingerprint, the spectral signature of a particular element is unique—each line in the fingerprint has a precise location, a very specific frequency at which it is always produced, whether in the lab or in the atmosphere of a star. What we detect, however, depends on the motion of the star. If the star is moving (either away from or toward us), these lines will be shifted: redward for receding stars, blueward for those headed our way. The spectral lines in the light we detect from stars wobbling back and forth under the influence of a planet will also shift back and forth in a regular pattern. The shift is exceedingly small, requiring extremely accurate measurements of the stellar spectrum. Thus, the search for extrasolar planets is limited to those in orbit about stars for which we can obtain very detailed spectra. Until very recently, we were restricted to stars at a distance of no more than about 600 light-years from Earth.

A second method for finding planets entails careful monitoring of the intensity of a star's light, looking for the barely discernible dimming that is caused by the transit of a planet across the face of a star. The planet blocks a tiny bit of the star's light each time it passes in front of the star, and astronomers look for a repeated, and regular, dip in the star's intensity. Fewer than a dozen planet transits have been observed, but because this method requires a high precision measurement of only the amount of light from the star, and not a shift in the star's spectrum, we can look for planets around more distant stars.

Prior to 2003, transiting planets had extended the reach of planet hunters to distances of 5,000 light-years, a new record.[10] That record was completely blown out of the water when the first microlensing planet was spotted in 2003. With this discovery we suddenly leapt 17,000 light-years into the Galaxy.

Microlensing Planets

A star creates a dimple in spacetime. A planet in orbit about that star adds a wart on top of the dimple. To find the wart you first have to find the dimple, then look very closely for the additional ding in spacetime produced by the planet. Blink and you'll miss it. A typical microlensing event caused by a Sun-like star in the disk of the Galaxy lasts a few months, but the spike in amplification caused by a planet is only a few hours to a few days long. So now we're looking for the scratch on the needle in the haystack.

To find the needle—the rare (one in a million) microlensing event caused by a star—astronomers have to cast a wide net, imaging millions of stars on a regular schedule, every few days or so. This schedule obviously won't work for finding planets. In order to document a spike that lasts only a few hours, images must be taken either continuously or at least every 10 minutes or so. Such an intense observational focus can be done only one event at a time. So astronomers have devised a profiling protocol—an alert system—for gravitational lenses.

As soon as the microlensing survey teams OGLE and MOA detect an increase in a star's brightness that is consistent with a microlensing event, they send an urgent message to a network of telescopes. These networks, such as PLANET (Probing Lensing Anomalies NETwork), MicroFUN (Microlensing Follow-Up Network), and even RoboNet (a prototype network of robotic telescopes) have access to telescopes all over the world, allowing the round-the-clock surveillance of a lensing event that is essential to finding a planet. By taking frequent images—on the order of minutes or hours—these telescopes can trace out the light curve in great detail, picking up any deviation from simple lensing.

The details are where all the fun is. When there is only one star

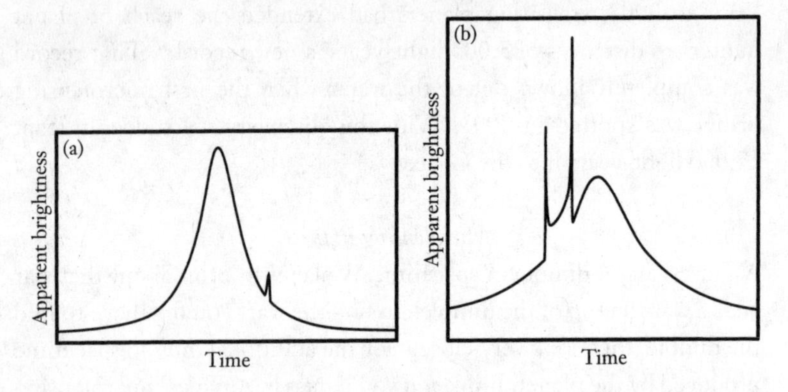

ILLUSTRATION 6.5 Finding planets. A planet in orbit about a star that is lensing an even more distant star will leave its imprint on the light curve—the apparent change in brightness of the distant star. The planet distorts the dimple in spacetime created by its star, and the observed effect depends on which part of this distorted lens passes in front of the distant star. (a) In this example the dimple created by the planet passes in front of the distant star. (b) Even more dramatic is the light curve produced when the very center of the lens passes in front of the distant star: the presence of the planet distorts the central portion of the lens created by the star, effectively etching lines of extreme magnification into this lens that create spikes in the apparent brightness of the distant star as they pass over it.

along the line of sight, there is only one lens. Add a planet in orbit about the star and you have two overlapping lenses. The deformation in spacetime is now much more complicated. In particular, some paths through this complex dimple in spacetime result in an almost infinite magnification of the light. What you see depends on which part of the lens you're looking through.

Two examples of a light curve produced by a lens composed of a star and its planet are shown in Illustration 6.5. Both represent a stark departure from the simple curve of a single stellar lens. The first example is fairly straightforward—the large magnification pattern is produced by the star, with a smaller (shorter-duration) spike due to the planet added on top. The second example is a bit stranger—there are two spikes, connected by a smooth U-shaped feature, imposed upon the underlying curve. In both examples the lens is composed of one star and one planet—but what we see depends on what part of this lens we are looking through.

To picture the lens, start with the dimple produced by a single stellar lens. The presence of a planet distorts this dimple, adding a second, much smaller dimple near the planet, but also disturbing the perfect symmetry of the larger lens created by the star. The smooth stellar lens is no longer perfectly smooth, but has two defects in it—one near the planet, the other near the very center of the lens.

If the part of the lens that is warped by the planet passes in front of a distant star, it will cause the single spike in amplification seen in Illustration 6.5a. The likelihood of this lineup is increased because of a fortuitous coincidence—the size of the Einstein ring for a stellar lens in the direction of the bulge is roughly the size of a typical orbit of a planet in our solar system. This increases the possibility that the planet will be positioned to magnify one of the images produced by the much larger stellar lens, creating the additional spike in magnification.

However, if the combination planet–star lens passes almost directly over the source star, so that the center of the lens is directly in front of the star, it will produce the double spike pattern seen in Illustration 6.5b. Because of the presence of the planet, a sharp line of extreme magnification (known as a *caustic*) traces around the very central region of the lens, etching a discontinuity into the otherwise smooth surface of the lens. The source star crosses behind the caustic as it nears the center part of the lens, then crosses behind it again as it exits the central region. Each spike in the light curve indicates that the sharp edge of the caustic has just passed in front of the source star.

These unique lensing signatures are produced by the same phenomenon that creates the dappled pattern in a swimming pool on a sunny day. As the water in the pool is jostled around, waves are created that bend the sunlight streaming into the pool. The overlapping waves and ripples in the water create effective lenses that at some points greatly (in theory, infinitely) magnify the sunlight. These caustics are clearly visible as you look into the pool, and in the gorgeous image in Illustration 6.6. Artist Brad Miller produced this photogram by shining a light through a vibrating tray of water onto photosensitive paper, creating a beautiful representation of a complex three-dimensional lens.

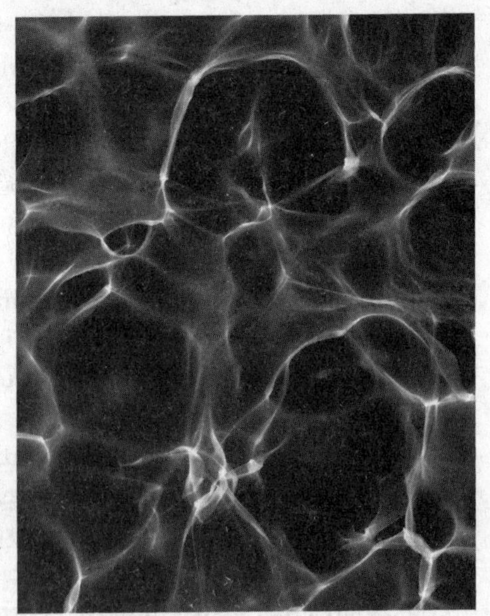

ILLUSTRATION 6.6 Lensing by water. Light shining through a vibrating tray of water onto photosensitive paper creates an intricate pattern similar to that seen on the bottom of a swimming pool on a sunny day. The light is bent and magnified by the ripples in the water—just as gravitational lensing light is bent and magnified by the ripples in spacetime.

New Worlds

The first microlensing planet was detected when the alert network traced out the characteristic double spike of a caustic crossing in great detail—an unmistakable sign of a combination lens. Weighing in at about one and a half times the mass of Jupiter, with an orbital radius about its star of approximately three times the distance between the Earth and the Sun, this new world completely shattered the previous record for farthest planet. OGLE 2003-BLG-235/MOA 2003-BLG-53 is 17,000 light-years away.

An additional twelve microlensing planets have since been discovered, with masses as low as 3.3 Earth masses and distances as far as 21,000 light-years away.[11] One has a mass in the Jupiter weight class, two are smaller than Neptune, and all are more than 9,000 light-years away. Included in this elite group is OGLE-2005-BLG-390Lb, whose detection in 2006 immediately set new records for both the smallest and the most distant extrasolar planet ever found. In orbit about a small red star (the most common type of star in the Galaxy), the mass and orbit of this new blue-ribbon planet place it in the same category as the giant ice planets in our own Solar System—Neptune and Uranus.

No other method would have been able to detect such a planet, even if its host star were much closer to Earth. Small sub-Neptune-sized planets far enough away from their star to avoid lethal doses of radiation simply don't have the necessary gravitational heft to be found via the Doppler technique. OGLE-2005-BLG-390Lb is still a bit on the cold side—scientists estimate its surface temperature to be a chilly 50 degrees Kelvin above absolute zero (–364 degrees Fahrenheit)—but it's the most Earth-like planet we've found so far. And only microlensing searches currently have the sensitivity to find more.

There is a downside to microlensing. We get only one glimpse of the planet—a single snapshot of the planet–star duo as it lenses a distant star. There is no opportunity to repeat the observation, and no hope of directly imaging the planet with a follow-up telescope observation to learn more about it, even though we know where to look. The planet is simply too far away.

What we can gain from these surveys is a better estimate of the numbers and masses of planets hidden within the Galaxy. If a microlensed star harbors a Jupiter-sized planet, the odds that we'll see it in the light curve are good. These odds drop precipitously (by a factor of 50 or so) for smaller planets, so the fact that half of the microlensing planets discovered so far are smaller than Neptune is a strong indication that these lightweight planets are much more common than their heavier cousins—an important result in developing models of planetary formation.

Microlensing planet searches are also opening the playing field to a wider group of observers. The alert search teams responsible for these new planets are not limited to large telescopes run by professional astronomers. The second microlensing planet was discovered with the help of two amateur astronomers in New Zealand who had signed on to the network of alert telescopes. Jennie McCormick, who obtained critical data on the caustic crossing with her 10-inch Meade telescope, summed it up this way: "It just shows that you can be a mother, you can work full-time, and you can still go out there and find planets."[12]

The Search for Another Earth

The future of planet hunting is very bright. In the fall of 2006, the Sagittarius Window Eclipsing Extrasolar Planet Search (SWEEPS) announced the detection of 16 extrasolar planet candidates. With the transit method, SWEEPS used the Hubble Space Telescope to look for planets in orbit about stars near the center of the galaxy, roughly 25,000 light-years away. The candidates exhibited short orbital periods (less than a day in some cases) and probable masses equal to or greater than that of Jupiter—further mass measurements are needed to confirm that these are indeed planets.

Within the next few decades NASA plans to greatly extend its search for extrasolar planets in a broad program entitled PlanetQuest. Several new missions will be launched, beginning with Kepler, a space-based telescope designed to search for Earth-sized planets via the transit method, which was successfully launched in March 2009. Next in line for the launchpad is SIM, which will target 50 to 100 nearby stars with higher precision than ever before in order to probe for terrestrial mass planets within the habitable zone—not too hot or too cold for liquid water to exist. SIM will also join the network of telescopes that respond to microlensing alerts, adding to the census of planets in the far reaches of the Galaxy and extending the range of planet hunters far beyond our local Galactic neighborhood. But the ultimate prize will lie within the sights of the Terrestrial Planet Finder (TPF), a future NASA mission that will use two complementary space-based observatories: one collecting visible light and the other an infrared interferometer. TPF will have the ability to hunt down Earth-like planets in the habitable zone—and to analyze the atmospheres of these extrasolar planets, seeking the biomarkers that signal the presence of life.

CHAPTER 7

Weighing the Universe

Hundreds of billions of galaxies are sprinkled throughout the visible Universe, but they are not evenly distributed across the vast expanses of space. Most galaxies are found in small groups of about 10, but a few percent reside in rich clusters of hundreds or even thousands of galaxies. A *cluster* contains galaxies of all sizes and shapes, whirling around each other in a cosmic game of mutual attraction. And just as the stars in the Milky Way are tethered to our Galaxy by an enormous sphere of dark matter, the galaxies in a cluster are bound to it by the overwhelming gravitational influence of matter we cannot see.

Clusters are the largest, most massive objects in the Universe, and most of their mass is in dark matter. The study of clusters is thus the study of the dark mass of the Universe. How much dark matter is anchoring the movements of the galaxies within a cluster—and what does this tell us about the dark matter fraction of the Universe as a whole? How is this dark matter distributed around and between the cluster galaxies, and what constraints does this distribution put on the fundamental nature of dark matter?

Gravitational lensing, with its power to trace out the dark matter in a cluster, offers a unique means of answering these questions. Because of their bulk, clusters create impressive gravitational lenses, carving out deep warps in spacetime that produce images of distant objects that are both visually compelling and scientifically of great importance—especially with respect to eliciting the recipe for the matter composition of the Universe. With these lenses scientists can weigh the Universe and begin to answer the question of how much and what kind of dark matter exists.

SAMPLING THE UNIVERSE

Seen through an optical telescope such as the Hubble Space Telescope, a cluster is a large collection of galaxies, with the brightest galaxy often positioned at roughly the center of the group. Observed with an X-ray telescope, a cluster looks like a big blob of hot gas. Viewed through Einstein's Telescope, a cluster appears to be a giant dent in spacetime caused by far more matter than is visible in gas and galaxies.

A typical cluster might be 10 million or 20 million light-years across and weigh as much as a quadrillion suns. It has three main components: galaxies, hot gas, and dark matter. The bulk of the cluster mass is in dark matter, and most of the normal mass in a cluster is in the form of very hot gas. This gas, with a temperature that is measured in tens of millions of degrees, shines brightly in X-rays. The cluster galaxies, scattered throughout the hot gas, appear to be a decorative afterthought.

Most important for dark matter estimates, a cluster encompasses a large enough section of the Universe that its composition—the relative amounts of normal matter and dark matter—are representative of the Universe as a whole.[1] This is not true for an individual galaxy, including our own. A galaxy does not represent a fair sample of the Universe—it's simply not a large enough piece of cosmic real estate to reflect the overall composition of the Universe. Using the relative amounts of dark matter and normal matter in the Milky Way to infer this ratio for the entire Universe would be like estimating the demographic makeup of the United States from the census data of a major city such as Chicago.

Clusters, on the other hand, do represent a fair sample of the Universe. Billions of years ago an enormous volume of matter, big enough to contain the universal ratio of dark to normal matter, began to collapse under its own weight in a process that would ultimately transform it into a cluster of galaxies. The matter in this volume was initially spread out fairly evenly, with only small variations in the density. These small variations grew under the influence of gravity, as regions that were slightly overdense to begin with attracted more matter, forming small clumps and strands of matter, which then merged to form larger clumps. As the clumps grew ever larger, the densities increased sufficiently that some of the normal matter collapsed even further to form stars and galaxies—a process not available to dark matter.[2] Thus, within the cluster that we observe today the dark matter and normal matter are clumped differently, with a fraction of the normal matter segregated into objects such as stars or the disk of a galaxy, while the dark matter is spread throughout the cluster. However, the average ratio of dark to normal matter in the cluster has not changed since the cluster first began to form—and thus this ratio still reflects the relative amounts of dark and normal matter in the Universe at large.

Fritz Zwicky was the first to plumb the cosmic information contained in clusters. His observations of cluster galaxies in the 1930s provided the first hints of the existence of dark matter, and for years, clusters were the only way to estimate the mass of the Universe. By calculating the fraction of normal matter in the cluster (relative to the total mass) and assuming that this fraction is the same for the entire Universe, we can determine the total mass density of the Universe.

It's a bit like taking a recipe for a casserole that serves six, and figuring out how many people you should be able to feed if you triple the amounts of each ingredient. Astronomers measure the total mass of the cluster, and the total normal mass of the cluster (the combined mass of the hot gas and the galaxies), and thus obtain the fraction of the matter in a cluster that is in the form of normal matter. The normal-mass density of Universe can be determined from either the primordial abundance of the light elements or the cosmic microwave background data, as discussed in Chapter 1. Because the relative fraction of the normal matter in a cluster is the same as that in the entire Universe, the total mass density of the Universe can found.

Finding the mass in the cluster galaxies is the easiest—a simple exercise in addition. Count up the galaxies in the cluster and sum their masses. The result makes it clear that galaxies don't carry much weight within the cluster. The amount of mass the galaxies bring to the table is relatively small compared to the other components—less than about 2% of the total mass in the biggest clusters.

Measuring the mass of the hot cluster gas is accomplished by observation of the X-ray intensity of the cluster. The temperature of the gas is so hot that it becomes ionized—electrons are stripped from their atoms, creating a soup of negatively charged electrons and positively charged ions known as a *plasma*. Collisions between electrons and ions in the plasma produce X-rays. The more electrons there are in the plasma, the more X-rays are produced; and since the number of electrons is related to the number of atoms of gas, the total gas mass can be determined from the intensity of the X-ray light.

Finding the total cluster mass is more challenging, for the obvious reason that most of it is dark. We have to extract the gravitational mass of the cluster, which we can't see directly, from something that we can see. There are three options—three categories of light sources: cluster galaxies, cluster gas, and distant galaxies. Each of these corresponds to a particular method for measuring the total mass of the cluster.

First on the list is the virial method, which relies on measuring the velocities of the galaxies in the cluster à la Zwicky. Cluster galaxies have been clocked at speeds of up to 2 million miles per hour in very massive clusters (using the same Doppler shift method used to determine the velocities of stars in the previous chapter), and these speeds are directly related to the mass of the cluster. Higher velocities imply more mass in the cluster. The underlying assumption in this method is that the cluster is reasonably stable—in a state of equilibrium. Many clusters are more or less middle-aged and relatively sedate. Though still accreting some mass at their outer boundaries, in general they aren't accumulating or losing great amounts of mass, and the galaxies within them should have had plenty of time to settle down to a speed more or less dictated by the cluster mass. However, not all clusters are this relaxed. If they've collided with another cluster, or a large galaxy has recently fallen in, the galaxies in the cluster may still be moving a bit more wildly in response to these disturbances and their motions will not be a reliable indicator of the cluster mass.

The second method also assumes that the cluster is in a fairly quiet state (equilibrium), but it uses the hot gas that surrounds the galaxies, instead of the galaxies themselves, to infer the mass of the cluster. The basic idea is similar to the balancing act described earlier for stars. Gravitational pressure pulling the mass of the gas inward toward the center of the cluster is balanced by the gas pressure outward—hot gas wants to expand (and cool); gravity acts to compress the gas (and thus heat it up). Equilibrium means that the two pressures are balanced. Thus if we can measure the gas pressure, we can determine the gravitational mass of the cluster. There are no cosmic pressure gauges, but the pressure can be found from a combination of the temperature and the density of the gas. (The pressure of an ideal gas is directly proportional to its density multiplied by its temperature.)

This is simple enough, as long as we have X-ray vision. Clusters emit prodigious amounts of X-ray light—light with energies 50 to 50,000 times higher than that of visible light. The temperature of the gas is encoded within the spectrum of this X-ray light. The X-ray spectrum of gas at a particular temperature has a characteristic shape, and by observing the intensity of the X-ray signal over a range of energies we can deduce the temperature.

There are two main obstacles to obtaining this X-ray information. First, the atmosphere of the Earth eats X-rays. In fact the atmosphere absorbs most light with energies above the ultraviolet (UV) regime, as well as a good chunk of the infrared light. Only visible light, a little bit of UV, and radio waves make it to the surface of the Earth, and even then the visible light is jiggled around by the molecules in the atmosphere, blurring the images we obtain with our terrestrial telescopes. We need to get up above the atmosphere to get a clear picture of the Universe. NASA, which has a penchant for putting things in orbit, has obliged with its four Great Observatories,[3] each targeting a different segment of the electromagnetic spectrum. The Hubble Space Telescope was first in 1990 and collects light centered on the visible part of the spectrum. The Compton Gamma Ray Observatory was sent up the following year to detect the highest energy gamma rays (it was "decommissioned"—sent on a controlled fatal plunge through the Earth's atmosphere—in 2000); the Chandra X-ray Observatory saw first light in 1999; and the Spitzer Space Telescope, tuned to light waves in the infrared, was launched in 2003.

Circling the Earth on an orbit that takes it almost a third of the way to the Moon, Chandra—named after Subrahmanyan Chandrasekhar, the 1983 Nobel laureate who worked out the details of stellar evolution—has improved our view of the X-ray Universe in the same way that the Hubble Space Telescope cleared up our vision in the optical, allowing us to see details that were previously obscured. The design of Chandra also addresses the second obstacle to X-ray collection. X-rays are very high-energy light, which means that instead of politely reflecting off a telescope mirror and into a detector, they burn right through glass or metal. The trick to capturing the X-rays is to avoid a direct hit. Thus, the mirrors on Chandra are shaped like barrels so that the incoming X-ray photons only graze the surface of the mirrors, skimming off the polished surfaces like a rock skipped across the surface of a lake, until they are collected by a detector at the bottom of the barrels.

Dozens of clusters have been imaged by Chandra with an increased spatial resolution that provides a detailed description of the temperature and density profile of each cluster. However, extracting the cluster mass from these data relies on an understanding of all of the details of the gas physics (which is complicated), as well as the assumption that the clusters are in equilibrium, which may or may not be true for a given cluster. Even more important, neither the hot gas nor the cluster galaxies necessarily trace the mass distribution in detail. Chandra data of several clusters have revealed that the gas is not always in a relaxed state of equilibrium, especially in the center regions of the cluster, and thus the X-ray profiles do not accurately trace the mass profiles of these systems. If we want to know how the mass is arranged within the cluster—a critical input for determining what the dark matter is—we need a cleaner measurement of the mass of the cluster, a probe that relies only on gravity. This leads us to the third method for finding the mass in a cluster of galaxies: gravitational lensing.

CLUSTER LENSING

Clusters are complicated and complex lenses. Each cluster is not one lens but many—one large lens fashioned out of the enormous dent in spacetime caused by the dark matter in the cluster, and many smaller lenses created by the individual galaxies within the cluster. Furthermore, clusters are built up over time, formed from multiple mergers of smaller conglomerations of dark and normal matter. Depending on how recently a cluster has collided with, merged with, or gorged itself on another system, the overall mass distribution in the cluster may not yet be thoroughly blended. The richness of the images produced by a cluster lens reflects this complexity.

The lens inscribed on space by a cluster is unlikely to be perfectly spherical, especially near the center, where one or two very massive galaxies can significantly distort the shape of the lens. Furthermore, the addition of many lenses multiplies the lensing effects that we saw in previous chapters. A planet–star combination lens creates caustics, which are essentially lines etched in the lens that produce extremely high magnification. A cluster lens will contain many more of these caustics—it is, in effect, a multifaceted gravitational lens.

The effects of such a complex lens can be striking. A cluster will lens anything that lies behind it, and it is so large that it always covers many light sources. Behind each cluster there are thousands of distant galaxies, millions or billions of light-years farther away from us than are the galaxies within the cluster, and occasionally a rare quasar. A quasar, which is essentially a point of light, can be lensed by a cluster into multiple images, so that we see not one but several copies of the same quasar. Galaxies are not points of lights but have a size and a shape that can be distorted by the cluster lens—a galaxy that lies behind a cluster lens will be stretched, magnified, and curved into a giant arc. This distorted image of a distant galaxy can also be multiplied many times over, creating a pattern of arcs that circle about the central region of the cluster.

The images produced by a cluster lens are thus obviously very different from the simple magnification of a background star seen in microlensing with its much simpler, and much weaker, lenses. Examples of such cluster lensing are shown in Color Illustrations 4 through 7. Furthermore, unlike microlensing by MACHOs or black holes within our Galaxy, cluster lensing does not produce an "event"—a time-dependent change in the brightness of a distant source of light.[4] A cluster lens is much farther away from us—too far for us to detect any motion of the cluster lens across the sky. Thus, a galaxy lensed by a cluster does not get brighter or dimmer as we watch.

However, a cluster is massive enough to produce a relatively large, and detectable, distortion of a background source. The multiple images seen in cluster lensing are separated on the sky by an angle thousands of times larger than those in microlensing (arc seconds, as opposed to the thousandths-of-an-arc-second separation of microlensing). From these images we can weigh the cluster, reconstruct its mass profile, test our models of the dark Universe, and uncover the most distant and primitive galaxies ever seen.

THROUGH A DARK LENS

The Universe of galaxies and quasars as seen through the gravitational lens of a cluster is both beautiful and surreal. The first glimpses of the strange effects produced by a cluster lens were serendipitous. A "filament-like structure"[5] was noted in observations of a large cluster (Abell 370) as early as 1976, but the possible connection to gravitational lensing was not made until almost a decade later, when new instruments came on line and clearer observations were possible. Even then, no one was looking for gravitational lensing when two groups of astronomers obtained detailed images of Abell 370 and another massive cluster, Abell 2218, in 1985. In fact, one group was testing its new cameras.

Near the center of each of these two clusters was an odd feature: a giant blue arc, over 300,000 light-years in length. Initial reports generated several possible interpretations for these odd configurations, from light echoes of extinct quasars to the bow shock of a large galaxy moving through the cluster gas, but it soon became apparent that these "rare and remarkable features"[6] were not part of the cluster, but distant galaxies, arched and magnified by the gravitational lensing power of the cluster.

Many cluster lenses have since been studied in great detail, and the Hubble Space Telescope has presented us with images of cluster lensing that should be appreciated first for their beauty, for the gorgeous distortions of distant quasars and galaxies that are clearly visible, and then for what they tell us about that which cannot be seen directly. Four clusters presented here (CL 0024+1654, SDSS J1004+4112, Abell 1689, and Abell 2218) beautifully illustrate the effects of cluster lensing and its power to reveal the dark matter content of the Universe.

Color images created from Hubble Space Telescope observations of the clusters are shown in Color Illustrations 4 through 7, and black-and-white annotated versions are presented in this chapter. In all of the images, the larger, bright yellow-orange galaxies are likely to belong to the cluster; the smaller and fainter galaxies that can be seen are not part of the cluster, but probably lie far behind the cluster, millions of light-years or more behind it. Astronomers determine exactly which galaxies are part of the cluster by measuring their redshifts (see the box on the next page). *Redshifts* are a measure of distance—groups of galaxies that are all at the same distance belong to the same cluster. In addition to galaxies, the images contain a few bright objects that have a symmetrical set of spikes radiating out from their center. These are either stars in our own Galaxy or quasars.[7] Everything else in the image, including all the faint little smudges, is likely to be a distant galaxy, unassociated with the cluster.

With these images, we can explore in detail the gravitational lensing power of clusters of galaxies. Four case studies of cluster lenses follow, each with unique features and each contributing to our understanding of dark matter.

REDSHIFT

Measuring the redshift of a distant object—a galaxy or quasar—is the standard method for estimating distances in cosmology. This method is based on a few key concepts:

1. Light changes color if the source of the light is moving toward you or away from you. The light will appear redder (shifted toward longer wavelengths) if the object is moving away; bluer (shifted to shorter wavelengths) if the object is moving closer. This phenomenon was discussed in Chapter 6 in describing one method for detecting extrasolar planets; it is also known as the *Doppler shift*.[8]

2. The Universe is expanding. The expansion of spacetime means that
   - All distant objects are moving away from us—and thus their light will be "redshifted."
   - The farther away an object is, the faster it's moving away—and thus the more the light will be redshifted.

⇒ *Objects with a higher redshift are farther away from us.*

3. The relationship between the speed with which a distant object is moving away and its distance is known as *Hubble's law*. The speed is equal to the distance multiplied by the *Hubble constant*, which is just the expansion rate of the Universe.

⇒ *If we know the expansion rate of the Universe and we measure the recession velocity of a distant galaxy, we can determine the distance to the galaxy.*

MEASURING REDSHIFT We measure redshift by looking at the spectrum of the light in detail, noting the pattern of dark lines that are caused by different elements in the light source. Hydrogen, calcium, and iron, for example, all imprint a unique pattern of lines on the light spectrum (see Illustration 6.4). These patterns consist of dark lines that appear at specific colors (or wavelengths), and are caused by well-understood physics of atoms. Each line always appears at exactly the same color (wavelength) if the light source is not moving. If the source of light is moving, however, the entire pattern of lines shifts toward the red or the blue end of the spectrum. The pattern itself—the spacing of the lines—does not change, allowing us to search for the pattern and determine how much it has shifted.

ILLUSTRATION 7.1 CL 0024 (black-and-white version of Color Illustration 4a). Five copies (A1–A5) of a single distant galaxy A are created by the lensing power of this cluster.

1. *Mapping Dark Matter*

One of the most striking cluster lenses is shown in Illustration 7.1 and Color Illustration 4a. The strange blue arcs circling about the bright cluster galaxies practically leap out of the picture. There are five copies of a single odd-looking galaxy. If the light of the cluster galaxies is carefully subtracted out of the picture, another three images of this same galaxy come into view.

Three of the images (A1, A2, A3) are highly magnified, but the central image (A5) is fainter than the original. The details of the underlying galaxy can be seen in each of the copies but, as expected in lensing, some of the images have been flipped by the lens just as the question mark was reversed in our earlier example (see Illustration 4.7c on page 85). Astronomers have been able to use these images to reconstruct the true shape of the lensed galaxy. It appears to be young and actively churning out new stars, with dusty regions that block out the starlight in patches so that in its original form the galaxy looks like a blue letter *B*.[9]

This reconstruction process also yields the details of the lens, in effect giving us a picture of the mass of the cluster. The lens details are extracted by an iterative process to create an increasingly precise map of the matter. It's essentially trial and error, using a computer: Make a model of the lens, put it in front of the distant galaxies, see what kind of image is produced and then compare this image with the real image. Repeat millions of times until you find the best match.

In creating the model lens, the best fit can be achieved by the adjustment of many parameters: the total mass of the lens, the shape and orientation of the large clump of cluster dark matter, the size of the flattened central section of the lens, the rate at which the mass decreases in density with distance from the center of the cluster, and the mass in each galaxy. It's a bit like adjusting the knobs on an LED projector until your PowerPoint presentation comes into sharp focus. Except, in the case of CL 0024 there is not one lens to focus, but 119—one enormous lens for the overall mass in the cluster, and 118 smaller lenses for the individual galaxies. Each of these lenses has several free parameters, and a model is defined by a specific set of these parameters. In all, over 2 million models were constructed (on a computer) and compared with the Hubble image.

The resulting "picture" of the lens is stunning. The total cluster mass within the region inscribed by the arcs is 237 trillion solar masses. The arrangement of this mass—the detailed structure of the lens—is shown in Color Illustration 4b. Each sharp peak corresponds to a galaxy within the cluster, and the larger mountain underneath corresponds to the bulk of the cluster—the enormous mass of dark matter in which the galaxies are embedded. Flip this image upside down and the connection with the rubber sheet analogy of Chapter 3 becomes obvious. The dark mass of the cluster creates a huge dent in spacetime, punctuated with the narrower dips due to the individual galaxies. This illustration represents the first high-resolution mapping of the dark matter in a cluster.

The profile of the dark matter portion of the lens is of great interest to cosmologists. The exact shape of this profile (which can be obtained by subtraction of the mass of the galaxies from the image) can give us hints to the nature of dark matter.

Most models for dark matter particles (e.g., WIMPs) predict that their density should increase fairly steeply in the center of the cluster. The mountain of dark matter in Color Illustration 4b should be sharply peaked. However, the observations do not appear to agree with this—the profile in the illustration (if you ignore the galaxy spikes) seems to rise rapidly from the outer edges of the cluster, then begins to level off near the center, creating a core region roughly 200,000 light-years in diameter—more like a softly rounded Appalachian mountaintop than the steep summit of a Matterhorn. There is an ongoing debate about whether this disagreement between theory and observation exists, and if so, how strong it is.[10] More data on the inner part of cluster halos from gravitational lensing will help to resolve this debate—and determine whether the behavior of dark matter particles in the form of WIMPs is consistent with the observed profiles of galaxy clusters. If not, we may need to consider even more exotic dark matter candidates.

2. *Quasar Quintuplets*

The Sloan Digital Sky Survey (SDSS) searched a quarter of the night sky for multiple images of lensed quasars. When they found SDSS J1004+4112, they hit the jackpot.[11] This cluster (see Illustration 7.2 and Color Illustration 5) has something for everyone. The cluster, which is about 7 billion light-years away, sits in front of a quasar that is roughly 3 billion light-years behind the cluster, and in front of a plethora of galaxies at even larger distances. Near the center of the cluster are four bright quasar images, and a fifth, fainter image of the same quasar is hidden in the glare of the bright cluster galaxy at the center. This is the first image of a quasar multiplied five times over by the gravitational lens of a cluster, and the widest separation of quasar images (15 arc seconds) ever seen at the time the image was obtained. Each quasar image lies on top of a faint smear of light, which is the lensed image of the galaxy in which the quasar resides. The galaxy images have been stretched into arcs, which trace out the Einstein ring of the cluster lens.

Three other galaxies have also been multiply imaged by the clus-

ILLUSTRATION 7.2 A bounty of lensing: cluster SDSS J1004 (black-and-white version of Color Illustration 5). Multiple images of a distant quasar (Q) and three background galaxies (A, B, and C) are highlighted. Five copies of the quasar (Q1–Q5) have been identified—the central image Q5 is almost hidden in the glare of the large galaxy at the center of the cluster. The faint smudge behind the quasar images is the lensed image of the galaxy in which the quasar resides. Two images each of galaxies B (B1 and B2) and C (C1 and C2) are seen, and five images of galaxy A have been circled. The fifth image (A5) stretches in the radial direction, pointing outward from the center of the cluster.

ter. There are five images each of one galaxy, and two images each of two other galaxies. Although most of the galaxy images are arcs circling about the center of the galaxy, the fifth image (A5) of the third galaxy is stretched instead in the radial direction, pointing outward from the center of the cluster like the spoke in a wheel. Such a radial "arc" can be produced if the mass distribution of the cluster levels off near the center and forms a core region[12] as discussed for the previous example (CL 0024), or if the central region is highly nonspherical (the center part of the lens is lopsided). We don't know yet which scenario best describes this lens.

The multiple arcs and quasar images seen in this Hubble snapshot are created by the complex lens structure of the cluster—the dimples in space produced by the overall (mostly dark) matter in the cluster and the individual galaxies distributed throughout it. To reconstruct the lens, scientists followed the iterative process outlined earlier, in the section on mapping dark matter. They first crafted a gravitational lens model that could produce the five quasar images as seen. Using this model lens as a first draft, they then looked for multiple images of other objects, and they found five images (A1, A2, A3, A4, and A5) of the distant galaxy A. They continued to refine the lens model until it could exactly reproduce these arcs.

The new and improved lens model, based on the images of the quasar and galaxy A, predicted the existence of high-magnification regions—caustics—within the lens. A focused search in the vicinity of these caustics revealed the presence of two additional lensed galaxies. Each is seen as a pair of images: (B1, B2) and (C1, C2).

Certain checks can be made to ensure that the lens model is correct. Whenever possible, redshifts are obtained for the various arcs. The model reconstruction process projects a distance to each of the lensed galaxies, and the measured distances of the images, obtained from these redshifts, are in good agreement with the model predictions. Further, the spectra of A1, A2, and A3 are identical, confirming that they are indeed multiple images of a single background galaxy.

From the final model of the lens, scientists calculated that the cluster weighs over 60 trillion solar masses.[13] Follow-up observations of the cluster with the Chandra X-ray Observatory (using the X-ray method described earlier in this chapter) provide a second, independent estimate of the mass of the cluster that is in excellent agreement with the value derived from gravitational lensing.

The quasar images contain even more information. The light emitted by a quasar is not constant, but changes in intensity over time in an irregular fashion. Thus, each image of the quasar will also vary, following the pattern of the light sent out by the quasar. But these images don't all vary in lockstep with one another. Careful observations reveal a time delay between the images. The delay is due to two factors, one geometric and one general-relativistic. The light in each image follows a different path through spacetime (see Illustration 7.3), and these paths are not of equal length—hence the geometric compo-

ILLUSTRATION 7.3 An artist's illustration of gravitational lensing of a quasar by a cluster of galaxies, producing a pattern similar to that seen in SDSS J1004 (Illustration 7.2). The light of the quasar is bent by the complex lens created by the cluster so that we see five copies of a single quasar.

nent of the time delay. The longer the path, the longer it takes light from the source to reach our telescopes. Different paths also take the light through different gravitational fields. Those that travel through highly curved spacetime (strong gravity) will be further delayed because the speed of light is slower (clocks run at a slower speed) in a strong gravity environment. This relativistic "Shapiro delay"[14] penalizes light that travels on paths close to the center of a mass concentration.

These two effects tend to work in competing directions as a function of distance from the center of the lens. As light races along multiple paths from the quasar to the Earth, a path that passes close to the center of the lens will go through a region of stronger gravity (thus experiencing a stronger Shapiro delay, which lengthens its travel time) than will a path farther out. On the other hand, a path farther from the center will be longer in distance. In many clusters, the gravitational delay slows the light traveling on paths close to the cluster center more than the extra distance slows the light on wide paths, so the light of the image farthest from the center arrives at our telescopes first, followed sequentially by the light from each of the other images as we move closer toward the center of the cluster.[15]

The pattern of changing light intensity seen in each quasar image is essentially identical but shifted in time. It's as if the quasar sends out a signal that we receive at four different times—first from one image, then echoed by each of the other images in turn. In SDSS J1004+4112, the four brightest quasar images were observed over a 3-year period, allowing the time delay between the two closest images to be determined, and constraints to be placed on the delays between the others. The light arrives via the image labeled Q1 first (see Illustration 7.2), followed 780 days later by the light from image Q2; 40 days after that the same pattern emerges in Q3, with a delay of over 1,250 days before it appears in Q4 (Q5 was not included in the analysis but its light would be expected to arrive last). Because these time delays are dictated by the mass distribution in the cluster, they aid in further refining our picture of the dark matter in SDSS J1004+4112.

3. The Incredible Hulk

Abell 1689, shown in Color Illustration 6, is one of the largest known gravitational lenses (which also makes it one of the largest lenses of any kind).[16] With its newly installed Advanced Camera for Surveys (ACS), the Hubble Space Telescope stared through this 2-million-light-year-wide lens for over 13 hours and was rewarded with the deepest ever view of the Universe. Revealed in this image are galaxies so distant and faint that they would have remained hidden from our view if the gravitational lens had not been there.

Dozens of arcs circle about the cluster center—long thin whispers of light that trace the curvature of space. Within this photo astronomers have identified over 100 arced images of 30 background galaxies—images that are found ever farther from the center of the cluster as the galaxies being lensed lie at increasingly large distances behind the cluster. Radial arcs can also be seen near the center of the cluster—small shafts of light pointing to the core of the mass distribution.

The mass profile of the cluster can be derived from this overabundance of cosmic stretch marks. The visible galaxies are spread out in several clumps, but the underlying dark matter distribution is remarkably smooth and symmetrical, creating a dent in the Universe with a mass of over a quadrillion times the mass of the Sun.

4. A Cosmic Magnifying Glass

In the previous three examples, the lens itself—the amount and distribution of the mass in the cluster—was the focus of scientific studies. However, a gravitational lens can also be a tool for studying other objects. Because lensing magnifies distant sources, a cluster lens can be used to bring into view objects that would otherwise be too faint to be detected. In particular, we want to get a closer look at the earliest galaxies. The details revealed by gravitational lensing of the most remote galaxies and quasars are essential for determining how galaxies formed, nailing down the link between quasars and black holes, and illuminating the end of the cosmic dark ages.

Abell 2218, one of the clusters in which the giant arcs of gravitational lensing were first seen, has proven to be an excellent magnifying glass. It was recently used by the Hubble Space Telescope to find the most distant galaxy ever seen, giving us a view of an early protogalaxy that could not have been detected without the lensing power of the cluster.

The search for these early galaxies requires that we understand the lens in detail. From the multiple arcs seen in Abell 2218 (shown in Illustration 7.4 and Color Illustration 7) it is possible to determine both the mass of the cluster—more than 60 trillion solar masses—and the details of its distribution. The model of the lens can then be studied, and in particular, areas of the lens where the magnification will be especially high—caustics—can be identified. These areas are caused by the composite nature of the lens, where the warping of space produces extremely strong magnification of light that hits just the right place on the lens. Once we know what the lens looks like, we know

ILLUSTRATION 7.4 Cosmic magnifying glass. Abell 2218 (black-and-white version of Color Illustration 7). This behemoth cluster of galaxies has created a giant lens in space that has produced multiple images of over a dozen separate galaxies, all of which lie far behind the cluster. Highlighted in this image are the multiple images associated with three of these background galaxies. (See text for details.)

where the caustics are—and we can search near them for highly magnified galaxies.[17]

Abell 2218 is not a simple cluster—instead of one large dent in space (plus lots of little dents from the galaxies), there are two. The main mass of dark matter is centered on the brightest galaxy, and most of the arcs in the image circle about this galaxy. There is a secondary clump of dark matter as well, positioned over the two bright galaxies near the top of the image, about two-thirds of the way from the left edge. Two distinctive arcs can be seen outlining this part of the dark matter lens.

Within this image scientists have found seven confirmed sets of images (and many more that have yet to be examined in detail)—each set consists of multiple copies of a single background galaxy.

- The most obvious arc is the reddest, marked as A1 and located at roughly two o'clock around the brightest galaxy. This arc corresponds to a galaxy located over 6 billion light-years away (corresponding to a redshift of 0.702), and another six images of this galaxy have been located about the cluster. Three of these images are marked A2, A3, and A4 (the others are hidden by the glare of cluster galaxies).

- The blue arc at about seven o'clock is actually two images (B1 and B2) of an even more distant galaxy, at 11 billion light-years (redshift 2.515). A third copy of this galaxy (B3) can be seen just outside the region claimed by the bright group of galaxies. It's small and blue, and it has a tiny white dot on it, a feature that can also be seen in its compatriots on the other side of the cluster.
- A third pair of arcs is almost too faint to be seen by eye in the Hubble picture—one of them is just barely visible as a tiny red smudge marked C1. C2 indicates the position of the second image, although it's too faint to be seen here. The source of these images is a galaxy at a redshift of 5.576, corresponding to a distance of 12.6 billion light-years—and at the time it was discovered it was the most distant galaxy ever seen.

The galaxy at 12.6 billion light-years was magnified over 30 times by the cluster lens and would not otherwise have been visible even in Hubble's long-exposure Deep Fields.[18] For a galaxy it's small—a few million times the mass of the Sun and only about 500 light-years across—and full of young, hot stars. It's most likely an early protogalaxy, a galactic building block that combined with others of its kind to form larger galaxies in a process of cosmic mergers and acquisitions that lasted billions of years. The stars within this protogalaxy are likely to be some of the first stars in the Universe—stars whose light ventured out into an otherwise dark Universe, ending the cosmic dark ages.

When astronomers homed in on the photos of this distant protogalaxy for a closer look, another, even fainter pair of images—of an even more distant galaxy—came into view. Like dinosaur fossil hunters who spend months carefully brushing away the dirt and rock around fragile fossils in order to extract the delicate remains, the scientists carefully examined the images of this faint galaxy in several different colors, combining the resources of the Hubble Space Telescope and the Keck Observatory telescopes (the largest optical and infrared telescopes, perched on the summit of Mauna Kea in Hawaii). The model of the lens of Abell 2218 also predicted that an even smaller third image of this galaxy should be found—and it was, exactly where it was expected to be.

On the basis of the data from Hubble and from Keck, the source of the images was determined to be a galaxy at a distance of at least 12.8 billion light-years (redshift 6.7). The lens has magnified this faint galaxy by a factor of 25, revealing some of its structure. Within the faint images the astronomers were able to pick out a bright central core, roughly 300 light-years across, along with a second, fainter knot of light—both indicating vigorous star formation, as intense as any seen in nearby galaxies today.

There are now several active research programs that peer through cluster lenses to search for the earliest galaxies. Using the Very Large Telescope, an array of four 8.20-meter telescopes and several smaller telescopes run by the European Southern Observatory, astronomers have identified over a dozen candidates for galaxies at distances that place them, along with the aforementioned galaxy, within the first billion years of the Universe.

Clusters have also been used as magnifying glasses to look more closely into the structure of distant quasars. The usual pointlike image of a quasar is blown up by the gravitational lens to reveal previously unseen structure, allowing astronomers to measure the disk of hot gas that is spiraling into the center of the quasar, and confirming the presence of a supermassive black hole at the heart of these energy giants. These new observations strengthen the link between distant quasars that blazed in the early Universe and the supermassive black holes that lurk at the centers of galaxies around us today.

The dark matter that weighs down a cluster creates a powerful gravitational lens. This lens produces giant arcs that circle about the center of the cluster and trace out the warps in spacetime created by the mass of the cluster—within the images produced by these lenses we are seeing the very real imprint of Einstein's theory of general relativity.

Einstein's Telescope can thus be used to effectively weigh a cluster and understand its true size and shape. What have we found so far? The data from observations of several cluster lenses (along with X-ray observations to measure the amount of hot gas they contain) provide clear results: dark matter dominates the mass of these clusters.[19] The mass in hot gas and galaxies—which are made of normal matter—accounts for at most 15% of the total; dark matter makes up the rest.

These results have striking implications for the composition of the Universe at large. Observations of the cosmic microwave background and primordial light element abundances imply that the average density of normal matter in the Universe is about 4% of everything—all the matter and energy that exist. If the ratio of dark to normal matter in the Universe is the same as that in clusters, there is almost six times more dark matter than normal matter. Roughly 23% of the total mass and energy in the Universe is in an entirely new kind of matter known only as dark matter—and composed of particles that have never yet been directly detected.

CHAPTER 8

Cold Dark Matter

Dark matter dominates the mass of the Universe, but the nature of this new form of matter remains unknown. The combination of all of the data we have available to us—the cosmic microwave background data, measurements of the primordial abundance of the light elements (nucleosynthesis), and the MACHO experiments—is definite in its insistence on what dark matter is not. Dark matter is not normal matter (including antimatter versions of normal matter), and it cannot be anything constructed out of our home team of quarks and electrons. None of these data, however, tell us what dark matter is. The challenge to answer this question has been laid squarely at the feet of both physicists and astronomers, and it will take an integrated effort by both of these groups to determine just what comprises the bulk (roughly 85%) of the mass in the Universe.

WIMPS

Theorists have done their part—in fact, they've been a little overenthusiastic. They have created a superfluity of possible dark matter particles, the most popular of which are the WIMPs mentioned earlier. Weighing in with masses on the order of 10 to a few thousand times the mass of a proton, and possessing a variety of interaction strengths (the measure of how likely they are to engage in a collision with another particle, such as a proton or another WIMP), WIMPs are hard to pin down, both observationally and theoretically. Over a dozen experiments are looking for WIMPs, but none have yet been seen. A few models have been ruled out by this nondetection, but many more remain out of reach of current experiments.

The generic requirements for a viable dark matter particle were laid down in the earliest moments of the Universe. In the era of particle soup, particles collided with one another at very high speeds—the higher the temperature of the Universe, the faster the particles zipped around. During these collisions particles of all kinds were produced, destroyed, and produced again. Every particle, including a WIMP, has an antimatter particle associated with it—a particle that is exactly the same in all respects except for its electric charge. When a particle and its antiparticle collide, they annihilate each other, and their mass and energy are transformed into pure energy, which then quickly decays into another pair of particles (one matter and one antimatter). The new particles can be different from the original two. Thus, a dark matter particle colliding with a dark matter antiparticle can annihilate and produce a pair of some other kind of particle. And vice versa; an electron and a positron, for example, can annihilate and create a dark matter particle plus a dark matter antiparticle.

In the very early Universe, these kinds of interactions go both ways with the same ease, even if the mass of the particles produced is much heavier than the mass of the annihilating pair. The interaction

WIMP + anti-WIMP → electron + positron

happens just as often as

electron + positron → WIMP + anti-WIMP

even though a WIMP is much heavier than an electron. The mass coming in is less than the mass going out, and this extra mass must come from somewhere. The answer lies in the high temperature of the Universe. Particles at high temperature are moving extremely fast, and there is energy in their motion. This energy of motion can be transformed into mass energy—the energy of the lighter particles is transformed into the mass of the heavier particles via $E = mc^2$.

As the temperature of the Universe drops, the motions of all of the particles decrease, and the interactions become one-sided. Dark matter particles can still annihilate when they run into their antiparticle twins, but they can no longer be produced by the annihilation of lighter particles. The energy of motion is insufficient to make up for the extra mass needed to produce the heavier dark matter particles. Thus, when the temperature of the Universe falls below a certain value, the number of dark matter particles also begins to drop precipitously—the birth rate falls far below the death rate.

If not for the expansion of the Universe, the number of dark matter particles would eventually plummet almost all the way to zero. As the Universe expands, however, the volume of space increases, and thus the density of all particles—including dark matter particles—decreases. The fall in density makes it harder for particles to find antiparticles with which to annihilate. At some point the likelihood that a dark matter particle will run into a dark matter antiparticle is, on average, essentially zero. When this happens, the number of dark matter particles is locked in—no more are produced or destroyed—a cosmological phenomenon known as *freeze-out*.[1]

Thus, the number of dark matter particles in the Universe today depends on the how many existed at freeze-out, which in turn depends on their interaction strength (how readily they annihilate with one another). Particles that interact strongly are more efficient at picking each other off. The stronger their interactions, the more annihilation will take place before the expansion of the Universe effectively puts a halt to this process, and thus strongly interacting particles will have lower final numbers than particles that interact more weakly. Therefore, broad constraints can be placed on dark matter models that propose to provide roughly the right amount of dark matter observed today. Their interactions can't be too strong or there wouldn't be enough particles left to make up most of the mass in the Universe. (Or too weak, which would result in too much dark matter.) In fact, calculations of the required strength show that it must be on the order of the weak nuclear force—hence the name *weakly interacting massive particles*, or *WIMPs*.

WIMPs belong to a more general category of dark matter that goes under the rubric of *cold dark matter* (*CDM*). CDM particles are cold because they're slow—at least compared to the speed of light—and they were already moving at nonrelativistic speeds when structures in the Universe began to form. This relative lack of speed allowed them to clump, adding their mass to any small lump of mass that they fell into. Think of the last time you played a game of miniature golf. As you putt the ball into the hole, it's important not to hit it too hard. You just want to tap the ball, so that it's rolling slowly when it reaches the hole; otherwise, if it's moving too fast, it will dip into the cup and right back out again. Likewise, matter particles must be moving relatively slowly in order to stay put inside a dip in spacetime (created by a clump of mass).

In a Universe whose matter is mostly CDM, this is the basic mechanism by which structures form—from the bottom up. Small clumps of matter form first and then merge with one another in a hierarchical process that creates ever-larger structures. Protogalaxies combine to form larger galaxies, and galaxies fall into clusters. The Hubble Space Telescope has recently spotted an early galaxy in the process of formation, capturing an image of dozens of small protogalaxies in the act of merging to create a larger galaxy.[2]

Scientists have also considered alternatives to cold dark matter, including faster-moving particles known as *hot dark matter*. Hot dark matter particles are usually lighter than cold dark matter candidates, and they don't slow down sufficiently to clump until much later in cosmic history. At the time when structures can first start to form (when matter becomes more plentiful than radiation), hot dark matter particles are still moving at near the speed of light, and thus they refuse to stay put in the small overdense regions that exist at this time. They view a shallow gravitational well as a speed dip, not a containing wall, and zip right over it. At the early times in the Universe when the more laid-back cold dark matter particles begin to build up the mass in regions with slight overdensities of matter, hot dark matter particles are too hyperactive to stay bunched together, and they stream out of these overdense areas, taking their mass along with them. A universe filled with hot dark matter will look very different from a universe dominated by cold dark matter at the same point in time after the Big Bang. Small overdensities are essentially erased in the hot dark matter universe, and structures much smaller than the Milky Way, such as diminutive dwarf galaxies, that form from these small overdense regions are stillborn.

Neutrinos are the quintessential hot dark matter candidate, and in the past they were high on the list of dark matter possibilities, in no small part because they have one key advantage over WIMPs: we know that neutrinos exist. However, current astronomical data have voted any kind of hot dark matter (as the main mass component of the Universe) off the dark matter island. The Universe we observe, with lots of smaller galaxies dotting the cosmic landscape, simply cannot be produced by hot dark matter particles.

Warm dark matter models have also been proposed, including a much heavier nonstandard model neutrino, whose velocities and structure-forming capabilities lie somewhere between cold and hot dark matter. However, cold dark matter remains the cosmological favorite, and WIMPs are the most popular variant of this genre.

WIMPs quite naturally arise in a theory, known as *supersymmetry*, that was first proposed in 1971 to address problems in particle physics unrelated to the question of dark matter. (This independence is considered a virtue that allows them to be viewed with less suspicion than other dark matter candidates.) Supersymmetry is part of the effort to find a deeper theory of everything.

Physicists love simplicity and symmetry, and Einstein was no exception. His final years were spent in quest of an underlying theory that would unify electromagnetism and gravity into one überforce and explain the entirety of the microcosmos with a single elegant theory. He was not successful, for reasons that include both the incomplete knowledge of the subatomic world at the time (the nuclear forces were not yet understood) and Einstein's own strong distaste for quantum mechanics.

A significant fraction of the available brainpower in theoretical physics today remains devoted to finding a unified theory. We now have an excellent quantum mechanical description of three of the four forces (electromagnetic, weak nuclear, and strong nuclear),[3] and a well-tested model that unites the first two. Including the strong force in a unified framework is a bit trickier, but we have many reasonable models that can accomplish this. Gravity remains the stubborn one. Our understanding of the subatomic world, and every measurement devised to study it, all agree that at the smallest scales the strange world of quantum mechanics reigns. Likewise, our view of the cosmos confirms that general relativity is an excellent theory for space and time over large scales. So what happens when we try to study space and time itself on the subatomic scale? In other words, is there a quantum mechanical theory of gravity similar to the quantum mechanics of the other forces? And if not, what new kind of physics describes the behavior of gravity—of space and time—in the subatomic realm?

We haven't yet convinced ourselves that we have the answer to this question, but we do have some promising leads. Supersymmetry seems to be a first step in the right direction. The concept of symmetry in physics is similar to our common use of this word. Saying that someone has a symmetrical face means that the left side is a mirror reflection of the right—both eyes are the same shape, size, and distance from the middle of the face, for example. Many symmetries guide our understanding of the physical world. Matter and antimatter illustrate a "charge" symmetry: the electric charge of a particle is different from the electric charge of its antiparticle, but everything else about the particle and the way it behaves is exactly the same.

Supersymmetry is another kind of charge symmetry, but to understand it we need to broaden our definition of charge. In a quantum mechanical view of the world, particles carry various kinds of charges that dictate how the particles behave in different situations. These charges are also known as *quantum numbers*. The most familiar quantum number is electric charge. Electrically charged particles have a quantum number (+1 for a proton, –1 for an electron) that determines how they act in an electromagnetic field.

The quantum numbers of a particle are ways of characterizing the particle, just as many different features are used to describe a person. Your driver's license probably lists height, weight, hair, and eye color, for example. This analogy works well in describing the quantum nature of charges. Charges cannot have just any value, but only certain discrete values. An electron has a charge of –1 exactly, and never –0.998 or –1.034. On a driver's license, only discrete values of height are listed: 5 feet 6 inches, for example, but never 5 feet 5¾ inches.

The charge, or quantum number, that is relevant to supersymmetry is *spin*.[4] The terminology is a bit confusing, since particles aren't really spinning in the same way that we think of the Earth spinning on its axis, or a baseball spinning on its way to the plate. In the Standard Model, particles are points in space and points, which have zero dimensions, can't spin. You need to be at least two-dimensional in order to define a point or an axis about which you spin. Furthermore, the spin of a particle can't be zero, nor can it be any random number. (Both of which are possible for a spinning ball.) Whenever we conduct an experiment to measure the "spin" of an electron, we find only two possible answers. The only spins it can have are up (+½) or down (–½).

The easiest way to think about spin is simply as another kind of charge that must be included in any quantum mechanical description of a particle. If we think of electric charge as height on a driver's license, then spin can be eye color. Like electric charge, spin is quantized—it comes in only certain discrete values. All normal matter particles, like the electron, have spin that comes only in units of ½ (the particle can have spin +½ or –½); all of the force carriers except gravitons have spin 1 (measurements of their spin always yield +1, 0, or –1). The graviton has spin 2, and some fundamental particles have been proposed[5] (but not yet seen) that would have spin 0.

This division of spin between matter and force particles raised scientific eyebrows. We don't expect drivers with blue eyes to be any better or worse at driving than are those with brown eyes. There should be a symmetry in driving with respect to eye color. We see such a symmetry with respect to electric charge: change the electric charge of an electron, and you get a positron, whose behavior in every other way is identical to the electron. So why not a symmetry with respect to the spin charge? Such a symmetry would posit that for every particle with spin 1 there should be a particle with spin ½ but identical in every other way; for every particle with spin ½, there should be one with spin 0. This is supersymmetry.

Supersymmetry assigns a new partner of a different spin charge to all of the existing particles—*selectrons* and *sneutrinos* are the spin 0 supersymmetric partners of electrons and neutrinos, and photons and gluons have their spin ½ *photino* and *gluino* counterparts. Among all these new particles could be one with just the right characteristics to be the dark matter particle. Physicists have explored the possible candidates for dark matter within supersymmetry and found the WIMP. It also goes by the less catchy name of *neutralino*—the lightest supersymmetric particle with spin ½ and electric charge 0.

Supersymmetry is also a component of string theory.[6] *String theory* mathematically models particles as vibrations on a string (as opposed to points). The details of string theory are still being worked out, and it's proving to be extremely difficult to connect the theory with experiments that can test it, yet it remains an area of very active research, and a current leading contender for a theory that can reconcile gravity with quantum mechanics—but we're not there yet.

String theory relies on supersymmetry to formulate its version of quantum gravity, but supersymmetry itself does not require string theory to be correct. Even if the ultimate theory of quantum gravity is not found in string theory, supersymmetry might still be a symmetry of our Universe. And unlike string theory, supersymmetry can be tested.

The problem is, no one has ever detected any of the supersymmetric partners predicted by supersymmetry,[7] so we don't know for sure that this theory is an accurate description of the real world. One of the major tasks of large particle accelerators, including Fermilab and the Large Hadron Collider, is to be on the lookout for signs of supersymmetry, including the production and detection of any supersymmetric particle. Should they find one, it will be hard to ignore the WIMPs' claim to the throne of dark matter.

WIMPs lead a roster of cold dark matter aspirants that also

include superheavy WIMPs (a.k.a. *wimpzillas*) and ultralight *axions*. Both are particle remnants of an earlier era of cosmic history. Axions are a fairly well-motivated dark matter candidate, and they were first introduced to resolve a problem in understanding the strong nuclear force. Thus, like WIMPs, they get points for being independently proposed in response to a separate problem unrelated to dark matter. But also like WIMPs, axion masses are not well constrained by theory, so that even if axions exist they may not be the dark matter. Wimpzillas are a more speculative proposition. If they exist, they were produced just after the Big Bang, and they outweigh WIMPs by a factor of more than a billion.

The list of dark matter contenders continues to grow as theorists run wild, unchecked by experimental data. There are two ways to rein them in: either find a dark matter particle and measure its properties, or go out and observe the collective behavior of dark matter particles in the Universe.

HUNTING DOWN DARK MATTER

Detection of a particle of dark matter is the realm of particle physics, and although dark matter doesn't make much of a gravitational dent in the Solar System, there should still be more than enough around us to make detection possible—the local density of WIMPs is about 15,000 particles per cubic meter. There are several ways to track one down: make one, capture one, or watch one self-destruct. Experiments anchored here on Earth are poised to create a dark matter particle in a high-energy particle accelerator; capture a particle of dark matter as it streams through Earth; or observe the flash of energy and particles given off when a dark matter particle and its antimatter twin self-annihilate somewhere in the Solar System or Galaxy.

Creation

The production of a dark matter particle is the purview of the large accelerators, such as the Large Hadron Collider[8] that was just built in the Alps near Geneva, Switzerland. Particle physicists are gathering from all over the world to construct and operate detectors for this particle accelerator, which will circulate two counterrotating beams of protons in a circular, underground tunnel almost 17 miles (27 kilometers) around. The beams will be collided, smashing one proton into another at energies many times higher than have ever been achieved before. Each proton will have an energy of 7 trillion electron volts (an electron volt is a standard unit for particle physics, representing the energy needed to move an electron across a 1-volt potential difference).

One of the main goals of the Large Hadron Collider is to search for supersymmetry. The proton–proton collisions will produce all kinds of particles, and because the energy of the collision is much higher than in previous accelerators, there is more energy with which to create new particles—and thus the potential to produce particles that are too heavy to have been seen before. The particle debris from each collision will be carefully examined to determine whether there are any new particles, including any of the predicted supersymmetry particles. With luck, we may find dark matter particles, but any supersymmetric particle that is found will confirm that supersymmetry exists. And if the existence of supersymmetry is confirmed, WIMPs are likely to exist as well.

Capture

The detection of a dark matter particle in the wild requires a different laboratory environment—often a deep underground retreat with hundreds of meters of rock to act as a shield against unwanted interference from cosmic ray particles. Cosmic rays are normal-matter particles from space that shower through the Earth's atmosphere in huge numbers and are studied in their own right as harbingers of various astrophysical and cosmological processes. But for dark matter hunters, they're a major nuisance. So dark matter detectors are placed deep underground—in a mine or under a mountain.[9]

Programs such as the Cryogenic Dark Matter Search (CDMS) set up shop at the bottom of out-of-service mine shafts, expanding the facilities already in place to accommodate the specific needs of their experiment. CDMS operates 2,400 feet under the surface of the Earth

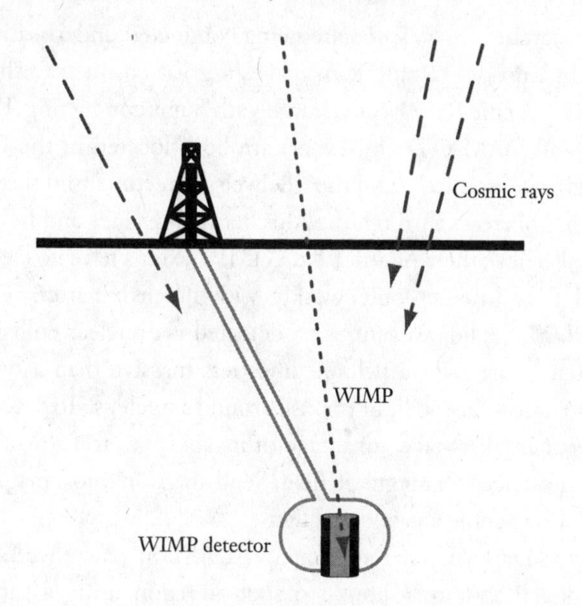

ILLUSTRATION 8.1 Underground WIMP detector. Experiments designed to look for WIMPs are often placed deep underground in order to keep normal-matter particles from space, known as cosmic rays, from entering the detector. Most cosmic rays, which cannot easily pass through the layers of rock and earth, will be stopped before they reach the detector. WIMPs, on the other hand, interact almost not at all with the normal-matter particles in the Earth, and sail through the rock as if it were not there.

in the Soudan Mine in northeastern Minnesota, placing its delicate detectors out of reach of cosmic rays, and shielding them as much as possible from radioactivity within the surrounding rock.

Even after cosmic rays have been dealt with, WIMPs are difficult to detect because they interact very weakly with normal matter—most of them fly right through the Earth and any detector set up to catch them. But every now and then a WIMP will bump into the nucleus of an atom in a detector, jiggling the nucleus ever so slightly as the WIMP glances off of it. This minicollision deposits a small amount of energy in the detector. In order to detect this energy, CDMS chills its detectors to just fractions of a degree above absolute zero, freezing the nuclei into place so that any detected movement is due to an external force—a collision with a WIMP—and not to the innate motions of the atomic particles within the detector.

Similar searches for WIMPs are being conducted underneath the mountains in Europe—a mountain makes a great cosmic ray shield. CRESST (Cryogenic Rare Event Search with Superconducting Thermometers) and DAMA (Dark Matter) are both located in the Gran Sasso tunnel in central Italy, and the Edelweiss detector sits in the Fréjus tunnel that burrows through the Alps between France and Italy.

Axions also have their Ahabs. Like WIMPs, axions have no electric charge and they interact only weakly with normal matter. These potential CDM candidates cannot be detected via nuclear collisions, however. Axions are over a trillion times less massive than a typical WIMP candidate—far too light to push around a nucleus. To detect an axion, experimentalists take advantage of its ability to transform into light in the presence of a magnetic field. Send an axion into a magnetic field and out may come a particle of light.

The Axion Dark Matter Experiment (ADMX) at Lawrence Livermore National Laboratory hopes to spot an axion using a tunable microwave collector—a hollow cavity with an adjustable size—immersed in a magnetic field. Scientists twiddle the tuning knob on the detector, changing the size of the cavity and thus its resonant frequency, searching for a signal just as you search for a station on the radio. When they hit the right frequency (when the resonant frequency of the cavity matches the axion mass), axions that are passing through will decay into microwave photons, signaling the presence of this elusive dark matter candidate.

There are also axion telescopes that look for axions produced in the magnetic field of the Sun, such as the CERN Axion Solar Telescope (CAST), a multinational effort currently in operation in Switzerland. Neither of these programs, nor any of their predecessors or contemporaries, have yet detected an axion, but they have begun to constrain the mass and interaction strength of this CDM candidate.

Self-Destruction

Some dark matter detection techniques are not looking for a dark matter particle, but for the products of dark matter self-annihilation. If the density of dark matter particles is high enough, a WIMP and an anti-

WIMP can find each other and go out in a blaze of high-energy particles, including gamma rays and neutrinos. To look for WIMP self-annihilation, scientists build telescopes that are sensitive to the gammas rays or the neutrinos and scan the sky looking for a burst of these particles to signal the presence of WIMPs.

The *average* density of WIMPs in the Universe (about 0.25×10^{-29} gram per cubic centimeter, which is equivalent to the density of one proton in a cube of space 3 feet on a side) is far too small for any significant amount of annihilation to occur, but WIMPs are not spread evenly through the cosmos. They gather wherever there is an overdensity of mass (and on large scales they are responsible for the overdensity). The density of the dark matter halo of the Milky Way in the vicinity of the Earth is about 400,000 higher than the average value in the Universe, and the halo itself is likely to be lumpy, allowing for regions of even higher density. The Sun and the Earth represent further opportunities for concentrating WIMPs. WIMPs can be captured by the Sun and Earth through collisions with nuclei in their interiors. These collisions are rare, but over the age of the Solar System a collection of WIMPs can be built up—and thus increase the likelihood of WIMP self-annihilation.

The products of WIMP self-destruction can be detected in gamma ray telescopes such as the Fermi Gamma-ray Space Telescope, which launched in 2008, or high-energy neutrino detectors such as AMANDA (the Antarctic Muon and Neutrino Detector Array). AMANDA and its next-generation successor IceCube are unique experiments, even by the standards of particle physics or cosmology.[10] The "detector" in these neutrino telescopes is a huge section of the pure ice underneath the South Pole. High-energy neutrinos streaming through the Earth occasionally interact with the nucleus of an oxygen atom in the ice, creating a muon in the process. The muon streams through the ice in the same general direction as its neutrino progenitor, generating light in its wake that is picked up by photosensors that have been dropped into holes drilled deep into the ice, extending over a mile beneath the surface.

AMANDA and IceCube also need to shield the ice detector from annoying cosmic rays, which can mimic the signal we're looking for from the neutrinos. It's a big detector (IceCube will utilize a cubic kilometer of ice), so a big shield is essential. The one that works best is the entire Earth. Almost all of the neutrinos from the WIMP self-annihilations pass through the Earth without stopping or interacting; cosmic rays do not. None of them will make it from the atmosphere above the North Pole all the way to the ice beneath the South Pole. Thus, any cosmic rays that enter the ice detector are coming from the atmosphere above the South Pole down into the detector (any other direction is blocked by the Earth itself), but the neutrinos can come from any direction (from above, from the side, or from below—from the center of the Earth) because they stream right through the Earth as if it weren't there. To avoid contamination of the data, the only signals that AMANDA or IceCube register are those from particles coming up into the ice detector from the direction of the center of the Earth.

The efforts to find a dark matter particle, one way or another, are scheduled to continue, as experiments are upgraded, refined, and improved. It's a challenging sport, but if dark matter particles exist, experimentalists are sure to get their hands on one sooner or later.

AN ALTERNATIVE TO DARK MATTER

Does dark matter even exist? The data presented so far are not in question—galactic rotation curves, cluster galaxy velocities, X-ray measurements, and gravitational lensing studies provide overwhelming support that something doesn't add up. But their interpretation as evidence for the existence of large amounts of unseen matter relies on one critical assumption: that we understand gravity.

What if this assumption is wrong? What if Einstein's grand theory of gravity—general relativity—is not valid in the regimes appropriate to galaxies and clusters? Does the rotation curve of our galaxy (and thousands of other galaxies) imply a huge amount of dark matter—or a new formulation of gravity? Putting it another way, does Einstein get the last word on gravity?

Modifying gravity—coming up with a new theory to displace or expand upon those of Newton and Einstein—is irresistible to theoretical physicists. It wouldn't be the first time gravity has been subjected to a major overhaul. We now understand Newton's laws of gravity as a special limiting case within the more general theory of Einstein's relativity. Under the appropriate circumstances, Newtonian physics provides a reasonable, and useful, approximation of the deeper theory. Newton will suffice for designing a roller coaster, but not a global positioning system. Newtonian gravity can explain, to high accuracy, the revolution of the Earth about the Sun—but not the orbit of Mercury. But is general relativity the ultimate theory? Perhaps Einstein's general relativity is embedded in an even deeper theory of gravity.

The current cosmological data make it abundantly clear that we need new physics. The question is, what kind of new physics? Einstein's equation (see Illustration 3.1 on page 54) has run into an observational obstacle. Something is missing from either the right-hand side of the equation, which describes the matter and energy in the Universe, or the left-hand side, which depends on gravity. Introducing new particles in the form of dark matter is one possible solution (a change to the right-hand side of the equation); messing with Einstein's gravity—adding something new to the left-hand side—is another.

There have been many suggestions for tampering with the gravity half of Einstein's equations, but current discussions of this topic usually start with MOND (Modified Newtonian Dynamics).[11] In 1983, Mordehai Milgrom proposed a modification to the force of gravity in order to explain galactic rotation curves without invoking dark matter. In Newton's formulation of gravity, the force between two massive objects decreases as the distance between them increases.[12] MOND posits that the gravitational force departs from this Newtonian prescription at large distances. In effect, at distances where the gravitational force is weak, MOND gives it an extra boost so that the strength of gravity doesn't fall off as quickly as it would under Newton.

Technically, MOND does not use distance as the marker that determines when to turn up the gravitational force; rather, it uses acceleration. When a force, gravitational or otherwise, acts upon a massive object, it causes that object to accelerate—speed up, slow down, or change direction. In the regimes where the acceleration due to gravity falls below a certain value (given by the parameter a_0, whose exact value is chosen to fit the galactic rotation curve data), MOND alters the gravitational force so that it doesn't decrease as quickly—it increases the strength of the force of gravity in the low-acceleration regime.

The distinction between distance and acceleration is important in the overall scheme of MOND, but in the case of stars orbiting about the center of a particular galaxy, acceleration and distance are directly related and so either one can be used. Gravity is modified (it's stronger then Newton would have it) on distances far from the galactic center (or equivalently accelerations below a_0). The net effect is that stars on the outer edges of a galaxy will orbit about the center of the galaxy at the same speed as those closer in. The rotation curve will be flat.

MOND does very well at explaining galactic rotation curves, but it is definitely not the answer to a new theory of gravity. Observationally, MOND fails to reproduce our observations of clusters of galaxies. The tweak to Newton's laws that works very nicely on galactic scales cannot explain away the need for huge amounts of dark matter in clusters. MOND must be supplemented with some kind of dark matter in order to be consistent with cluster data, and massive neutrinos (the hot dark matter described earlier) are often invoked to play this role. Which makes MOND less appealing—it requires both the modification of gravity and some kind of dark matter.

More important, however, MOND is not a theory, because it doesn't posit the fundamental physics that bring about these changes in the laws of gravity. It simply describes what the new laws should be in order to fit the data. This doesn't mean that there is no underlying theory that would result in such an empirical alteration to the laws of gravity—but MOND itself is not such a theory. And without a full relativistic theory, MOND cannot be put to the necessary test of comparison with all of the available data, such as the cosmic microwave background, the observations of structure in the Universe (the numbers of galaxies and clusters, and their distribution), the expansion history of the Universe, and gravitational lensing observations. Predictions for these data cannot be derived from an empirical formulation.

This is critical. Any suggestion for new physics must first satisfy all the existing data (and there's an abundant amount); then, to be taken seriously, it must also make new predictions that can be tested with experiments. Two competing theories, proposed to explain data already in hand, will often make different predictions for future observations. The proponents of new theories, as well as the critics, are obliged to outline the ways in which their theories can be differentiated from the rest of the pack.

Nonetheless, there are some limits to a Universe with MOND and no cold dark matter. Even if we toss in the massive neutrinos required to reconcile MOND with cluster data, the bulk of the matter in a MOND scenario is normal matter. In MOND, the sources of gravity are the gas and stars that are visible. The effects of this new gravity *must* trace the observed distribution of normal matter. This requirement provides a crucial test for modified gravity models.

Although MOND is not a full-fledged theory, it could be a starting point. Many attempts have been made to develop a theory of gravity that reduces to MOND on the scale of galaxies, and gravitational lensing observations have played a large role in directing these efforts. Gravitational lensing is a general-relativistic phenomenon, which arises from the curvature of spacetime by mass. MOND addresses only the limited Newtonian regime of gravity and thus has nothing to say about lensing. However, it's clear that a full theory of MONDian gravity must include some way to produce strong gravitational lenses from the small amounts of normal matter that we see in clusters and galaxies. The observations of cluster lensing that we saw in the previous chapter require much more mass than the normal matter in the cluster. Because MOND has only normal matter, a broader theory must include a different mechanism for reproducing the multiple images and giant arcs that are clearly seen in lensing—and many early attempts at a full theory of modified gravity were discarded because they failed to meet this challenge. A new theory of gravity has got to do it all.

The most promising alternative theory of gravity so far was put forth by Jacob Bekenstein in 2004: the Tensor-Vector-Scalar (TeVeS) theory.[13] The description of this theory is complicated, but the main point is that Bekenstein added two new players to the game of gravity. These new players change gravity in such a way that our simple mental picture of bowling balls on a rubber sheet no longer works. The response of spacetime to mass and energy is not so simple.

In general relativity, spacetime is represented by a field called the *metric tensor*. This is a technical name for an entity that is essentially a dynamic and flexible yardstick—one that describes the curvature of spacetime at every point in the Universe. Matter responds to the metric (the curvature of space in a particular location) and also influences the metric (where there's more matter, there's more curvature). A bigger mass induces a correspondingly bigger dent in spacetime.

TeVeS complicates this simple connection by the addition of the two new fields. As you might guess from the name of the theory, one field is a scalar and one is a vector—the tensor is the same metric as before. A scalar field is something that has a value at every point in space (like the depth of water in a lake); a vector has a value and a direction (like the flow of water in a river).

Matter now responds to a new metric, which gets contributions from the old metric—the curvature of spacetime—plus the two new fields. The strength of these contributions varies depending on the circumstances, so TeVeS looks just like general relativity when gravity is strong—near a black hole, for example—and like MOND when we calculate the rotation curve of a galaxy. The scalar field is a long-range force field (in a sense, a new "fifth force") that changes the strength of gravity, and the vector field adds some extra bending of light in order to induce sufficient gravitational lensing with less matter. The bottom line is that the way matter and spacetime relate to one another is no longer always face-to-face. The honeymoon is over and the interactions between the husband and wife are now also influenced by the neighbors and the in-laws.

We can use several key features of TeVeS to sum up this detour into theoretical physics. For high speeds and large accelerations, TeVeS reduces to general relativity. For low speeds and small accelerations (such as those we experience on Earth), it looks just like Newtonian gravity. And at even lower accelerations, it morphs into MOND.

TeVeS can also make predictions for cosmological observations. Not surprisingly, it's a complex theory, and extracting these predictions is painful. But recent efforts to beat some predictions out of TeVeS have yielded answers. It does relatively well up against most of the cosmological data, but only if two other major components—neutrinos and dark energy—are included along with normal matter in the cosmic inventory. The bulk of the Universe in this model is in dark energy (78%), while the neutrino contribution is ratcheted up to about 17%. (CDM models put an upper limit of about 3% in neutrinos in order to avoid wiping out too much small structure.)[14] Normal matter comes in a distant third, at 5%.

At least two tests for TeVeS remain. The first comes from the cosmic microwave background. The predictions for the detailed pattern of the CMB are slightly different in a Universe described by TeVeS with dark energy and more mass in neutrinos, than in a Universe with Einstein's gravity, cold dark matter, and dark energy. The observations are not yet precise enough to distinguish between these two models, but new data from the Planck satellite, which launched in May 2009, should settle the question.

The second test requires us to get back into the thick of cluster lensing. If gravity is modified and dark matter does not exist, the gravitational imprint of the mass in a cluster must always trace the visible mass in the cluster gas and galaxies. Lensing sees only mass, and if normal matter is all there is, then lensing observations must be centered on the normal mass we can see. If dark matter does exist, this restriction no longer holds. The dark matter and the normal matter may not always be in exactly the same place. By revealing the distribution of matter throughout the cosmos, lensing can help to uncover the true nature of the matter in the Universe, and determine whether Einstein's gravity still holds.

CHAPTER 9

Tracing the Invisible—and Finding Dark Matter

While particle physicists are busy collecting and dissecting subatomic specimens of interest, astronomers are tackling the problem from a different perspective. They are the field biologists of the dark matter team, charged with observing the behavior of various dark matter populations in their native habitats. Their job is to find dark matter in the Universe at large, and map its distribution—to document how much dark matter exists, and how it arranges itself in and around galaxies and clusters. These field studies can test alternative models of gravity and determine whether Einstein's theory of general relativity needs to be modified. They can narrow the range of viable dark matter models, thereby allowing the particle hunters to focus their resources on detection techniques aimed at specific types of dark matter. And as an added bonus, the more astronomers find out about dark matter and its proclivities, the more tightly these observations can be used to constrain rogue theorists—or inspire them to new levels of creativity.

COLD DARK PROBLEMS

Cold dark matter models—models of the Universe with 23% cold dark matter, 72% dark energy, and 5% normal matter (scientists refer to this model as ΛCDM, where Λ is the Greek lambda, the dark energy component)—are in good agreement with many recent observations of the Universe. These observations are made on different scales, using different kinds of experiments that probe the Universe at many different times in its history. These data, such as the cosmic microwave background, galaxy and cluster mass determinations, and the general picture of structure formation all support the basic picture of a Universe filled with cold dark matter (and dark energy).

However, questions still remain: Is cold dark matter the answer to the missing mass of the Universe? Or is it instead another kind of dark matter? Or another kind of gravity? To answer these questions we need to take a closer look at dark matter halos.

A dark matter halo is essentially a cloud of dark matter. Halos come in a wide range of sizes—small halos around the smallest galaxies, larger halos around larger galaxies, and gigantic halos around large clusters—but models of cold dark matter predict that they should all have the same basic shape and profile. They are roughly spherical (although probably not exactly), with a density that is highest in the center. The density of the dark matter keeps dropping as you move out from the center of the halo and, like a cloud, the edges are not well defined. It's hard to say exactly where a halo ends, and the halos of two galaxies that are near each other often meld together at some point.

There are dark matter halos around galaxies, much larger halos around clusters of galaxies—and possibly dark matter halos that contain no visible matter at all. Any dark matter halo should also contain multiple smaller "subhalo" objects—smaller clumps of stars, little (dwarf) satellite galaxies, and clumps of dark matter in which no stars were ever able to form.

Halos of whatever size have been assembled over time by gravity. Even the largest began as tiny overdense regions in the very early Universe that attracted more and more mass as the Universe evolved. Exactly what these halos should look like today depends on the nature of the missing mass. The more precisely we can map out dark matter halos, the more we will know about dark matter and its alternatives.

The process for testing cold dark matter models and alternative gravity models with dark matter halos involves several steps, which are usually overseen by different groups of scientists. Particle physicists calculate the relevant characteristics of their favorite dark matter particle—its mass and how it interacts; numerical cosmologists input this information into computer simulations of halo formation; and astronomers go out and take the measurements of real halos to determine their size, shape, profile, location, and subhalo content. The observed halos are then compared to the simulated halos to find out which models give the best fit.

So far, observations and predictions don't always agree. Models of cold dark matter (CDM), such as WIMPs, predict that the center of a dark matter halo should be sharply peaked—the density in the innermost regions of a dark halo should rise very steeply toward the center in a *cusp*. CDM models also predict that, in addition to the galaxy- and cluster-sized dark matter halos, many more small clumps of dark matter should exist. Dark halos are formed by the infall of lots of smaller objects, some of which get shredded and mixed into the overall halo, but some of which should still be more or less intact.

Observations paint a somewhat different picture of halos. Galaxies and clusters are observed to have central *cores* instead of cusps—the density of the dark matter halo appears to level off within the central region. And not enough small structures have been seen. CDM predicts that there should be several hundred small "dwarf" galaxies orbiting through the dark halo of the Milky Way. Only about a dozen have been found.

However, it's not yet clear that there is a conflict. The computer simulations are difficult and keep pushing the limits of our computing power; and the astronomical observations are also challenging, especially in obtaining the small details that are crucial for probing the nature of the dark matter.

Sim-Universe

The computer simulations that create dark halos from tracking the behavior of CDM under the influence of gravity have become increasingly precise, but the numerical challenges involved are still significant.[1] The basic idea is fairly straightforward: Start with an enormous number of dark matter particles distributed throughout a three-dimensional box that represents a chunk of the Universe. Calculate the force on each particle from all of the other particles. Take one extremely tiny step forward in time and reposition each particle according to where this force would have it move.

The challenge lies in producing an accurate simulation of both the larger picture (the formation of the entire halo over billions of years) and the smaller features (the exact shape of the halo in the central region or the fate of tiny subhalos). To be accurate over small distances (the size of halo centers), simulations need to include more dark matter particles and shorter time steps. They also need to track the normal matter in the halo, which can clump more tightly than the dark matter (in stars, for example) and may thus gravitationally affect the behavior of the dark matter on small scales. Normal matter necessitates encoding the physics of gas and stars, and it requires detailed, focused calculations in very small regions of space.

On the other hand, following the behavior of the dark matter in a sufficiently large chunk of space (one that encompasses many halos and their effects on one another) over almost the entire age of the Universe is possible only if fewer particles of dark matter are tracked and the time steps are larger.

Only so much computing power is available, and deciding how to allot it in order to get both the big picture and the minute details simultaneously is difficult. Taking a group photo of 300 people that also allows you to see the fine detail in each person's face requires an impressive camera and an expert photographer—and there is still a limit on how well it can be done, given current technology.

Observational Challenges

Astronomical observations of the inner regions of galaxies and clusters are also challenging—there's much more action in the center than farther out. The gas and stars in the inner portion of the galaxies and clusters are frequently stirred up and less likely to have settled down to an equilibrium state, so it's difficult to use them to trace out the mass in these areas. The central regions are also relatively small—less than a few thousand light-years across in a galactic halo that extends to hundreds of thousands of light-years, or less than a few hundred thousand light-years in cluster halos whose overall size is on the order of 10 million or 20 million light-years—and thus harder to see.

Finding subhalos is equally challenging—they may be completely dark (star-free) or filled with only a few faint stars that are lost against the background of other stars within the galaxy or cluster. New dwarf galaxies (too faint to have been detected by earlier searches) continue to be found in the halo of the Milky Way,[2] suggesting that we shouldn't stop counting yet. Furthermore, many of the small subhalos expected in large numbers in a galaxy halo may have been stopped short in the process of formation. The birth and early death of massive stars in small subhalos result in catastrophic supernova explosions, which can produce strong winds that empty the subhalo of its gas, and thus halt the formation of any more stars.

If both the simulations and the observations hold up, however, the discrepancies between the two may be telling us something about the nature of dark matter. If a simulation uses a dark matter particle whose interactions are not those of a typical WIMP, the halos that are created in the simulation will be different. In standard cold dark matter models, cores (as opposed to cusps) do not form in a galaxy or cluster halo—but they might if the dark matter particle has more complicated interactions with itself that prevent its density from getting too high. The lack of subhalos could point to the influence of a neutrino hot dark matter contribution to the dark matter inventory (although cold dark matter would remain the major component). As discussed in the previous chapter, neutrinos stream out of small overdensities, thereby wiping out all traces of structures below a certain size. Subhalos might never form if there is enough neutrino dark matter around.

While the cosmologists continue to improve their simulations and particle physicists postulate new kinds of dark matter particles, Einstein's Telescope holds the best hope for improving the observational piece of the dark matter puzzle. To determine whether cold dark matter is the answer to the missing mass in the Universe, or instead something else—another kind of dark matter or another kind of gravity—is required, we need to look closely at dark halos with a more subtle form of gravitational lensing.

STRONG VERSUS WEAK LENSING

The dramatic multiple images and arcs created by large cluster lenses are relatively rare. They are the result of *strong lensing*—lensing that is produced when a light source lies behind a very pronounced warp in spacetime, such as the center region of a cluster, where the concentration of mass is very high. The multiple images and Einstein rings created by the lensing power of individual galaxies, and the amplification of starlight due to microlensing by MACHOs (which produces multiple images even though we cannot resolve them), are also examples of strong lensing.

Most lensing is much more subtle. Smaller mass concentrations, which create shallower dimples in spacetime, or situations in which the sources lie farther out from the center of the lens (beyond the region of high density) will produce weaker lensing signals. Such *weak lensing* occurs around every massive object—every galaxy, every cluster, every dark strand of matter in the Universe. The effects of weak lensing are hard to detect. No multiple images or giant arcs are produced; instead, the image of the source is altered in ways that are barely perceptible. Weak lensing changes the position, shape, and brightness of a source by a tiny amount.

The very first detection of gravitational lensing—the observed shift in the position of stars during the 1919 solar eclipse, which was discussed in Chapter 4—falls into the category of weak lensing. The stars observed by Eddington and his colleagues were shifted in position by only a very small amount, and no double images were produced.

Weak lensing on the cosmological scales relevant to the search for dark matter is even more understated. The effects on any one distant object—a galaxy or quasar—are effectively impossible to detect. There is no way to know exactly what the (unlensed) source looks like, or exactly where it is. The solar eclipse measurements depended on the ability to measure the positions of the stars when the Sun (the lens) was not in front of them. We don't have this option for more distant lenses—on any human timescale, a galaxy lens doesn't move.

ILLUSTRATION 9.1 Weak lensing of a circular source of light. The shape of the light source is distorted to become slightly elliptical, with the long axis of the ellipse oriented tangentially about the lens.

Over the past 10 years or so, however, weak lensing has rapidly emerged as one of the most valuable tools in the search for both dark matter and dark energy. The key to detecting the minute distortions caused by weak lensing is in the numbers. Small effects, unnoticeable on their own, can make a strong statement when added together. A penny is not worth much today, yet if all of the now more than 300 million people in the United States were each to put just one penny in your bank account, your wealth would increase by over 3 million dollars.

Weak lensing gives us penny-sized bits of information—but these bits add up to a wealth of data. Everything behind the lens is affected by its presence. A galaxy lens might distort our view of 10 or 20 distant galaxies, while behind a cluster lens lurk thousands of galaxies billions of light-years farther away. It is the collective behavior of the galaxies seen through a lens that allows us to "see" the lens.

The type of change induced by weak lensing is shown in Illustration 9.1. Suppose a distant light source is a perfect circular disk. As the light from this disk passes through a gravitational lens, it will be deformed into an ellipse, with its long axis oriented roughly tangentially about the center of the lens. (The lensed shape of a galaxy is technically more of a crescent—a tiny arc, or "arclet"—but the distortion is so small that an ellipse provides a reasonable description.) This effect is known as *shear*.

The effects are greatly exaggerated in this cartoon—in reality, the average change in the ellipticity of one galaxy lensed by another is less than 1%. Furthermore, real galaxies are not round. A galaxy is already roughly elliptical in shape, and weak lensing will simply make it more or less so, reorienting the long axis of the ellipse in the process. To further complicate the situation, galaxies come in a variety of shapes—some almost spherical, others highly elliptical. Since we don't know the original shape or orientation of any individual galaxy, we have no way to determine how much it has been distorted by lensing.

However, weak lensing will change the average appearance of the galaxies that it affects. Weak-lensing studies are based on statistical analyses of the shapes and orientations of the background galaxies. By observing the effects of weak lensing on large numbers of galaxies, we can tease small changes in the average shape and orientation out of the data.

First, astronomers must take into account the fact that galaxies have a range of intrinsic ellipticities. They do this by averaging the shapes of thousands of galaxies, especially those that do not lie behind a cluster or another galaxy. The resulting average provides a benchmark shape for comparison to the average shape of galaxies behind a lens.

Next, add in the assumption that the orientations of distant galaxies are completely random.[3] Toss a large handful of jelly beans up into the air and, if they land on a smooth floor, they will randomly point in all directions. We expect galaxies to be sprinkled across the sky in a similar way. Lensing looks for a departure from this randomness. The galaxies that lie behind a lens galaxy will, on average, tend to point a little more in the direction tangential to the center of the lens than not—the lens orientations will no longer be completely random.

The change in the average shape and orientation of lensed galaxies will still be very small, and how well we can map the dark lens will depend on how many galaxies we can see behind it. The widely different numbers of background sources for weak lensing by a galaxy or a cluster force us to analyze the observations for these two kinds of lenses in a somewhat different fashion, but both offer a window into dark matter halos.

DARK MATTER HALOS, PART I: GALAXIES

Each galaxy in our Universe is embedded in an enormous sphere, or halo, of dark matter. If, with our eyes or our telescopes, we could "see" this dark matter, what would galaxies really look like? Is the dark halo truly spherical, or is it somewhat flattened? How big is it?

When we look through a galaxy lens, we can usually see about 10 distant galaxies whose shapes have been slightly distorted by the lens. None of these galaxies by themselves look strange, but their average orientation and shape will be altered as described already. However, 10 is still not a very large number—not nearly large enough to allow us to map out the dark halo of the lens galaxy. This means that we can't find the size and shape of any individual galaxy dark matter halo from weak-lensing observations. We can, however, determine the characteristics of a typical dark matter halo. Just as doctors conduct studies involving hundreds or thousands of patients to determine how a new treatment affects an average patient, astronomers look at a huge number of galactic lenses to create a picture of an average dark matter halo.

To obtain the large numbers of distorted background galaxies that are needed, scientists collect photos of many lens galaxies through which 10 or so background galaxies can be seen. These photos are divided into categories based on the lens galaxy—those that are similar in mass and shape are lumped together. The images of many lens galaxies are then stacked—images of similar galaxies are laid on top of one another to form one composite, "average" galaxy, which now has hundreds or thousands or more galaxies behind it (the total of all the background galaxies of all of the stacked lenses). The small shear effect, while not visible for any single lens galaxy, becomes more and more noticeable as more galaxies are added to the stack. It's almost like watching the creation of a picture made out of dots. With 3 dots you have no idea what the image will be; with 100 dots a hint of an outline may appear; by the time 1,000 dots have been printed, the shape of a human face becomes visible; and with millions of dots you can identify the person.

Only recently have such studies become possible. The first observations of weak lensing of galaxies were obtained in 1996, in a study that examined 439 lens galaxies and 506 fainter source galaxies.[4] Since then, surveys of millions of galaxies, such as the Sloan Digital Sky Survey (SDSS), have gotten into the act. The measurements are difficult, limited by our ability to make out the shapes of faint galaxies, but the results are clear. Einstein's Telescope has given us the first composite portrait of the dark matter halos that surround and dominate galaxies.

The Sloan Digital Sky Survey was originally designed to create a three-dimensional map of a huge chunk of the Universe. Astronomers have been charting the heavens for thousands of years, but the early maps were all in two dimensions. When you look up at the night sky, or take a photograph of what you observe through a telescope, the resulting image is a two-dimensional projection of a three-dimensional Universe. Two galaxies that appear right beside each other in the photo may actually be separated by billions of light-years—one may be much closer to us than the other. Adding in the third dimension is not easy. The distance to each astronomical object must be determined, which for galaxies means measuring the redshift of the light in their spectra (see Chapter 7). Obtaining spectra is time-consuming, but SDSS is able to significantly speed up the process by collecting the spectra of 600 objects simultaneously. In the summer of 2008, the survey completed its target goal of obtaining redshifts to about 1 million galaxies, and with these data over one-fourth of the sky—more than 10,000 square degrees—has been charted, creating the largest map ever made.

In addition to targeting galaxies for spectral observations, SDSS is cataloging optical images of nearly 200 million objects—stars, quasars, and galaxies. This enormous database is just what is needed to trace out the dark matter halos of galaxies using weak lensing. In 2004, members of the SDSS collaboration reported the results of a weak-lensing study of galaxy halos in which they looked at 127,001 lens galaxies and 9,020,388 source galaxies. By 2006, another group within SDSS had upped the ante significantly, peering through over 2 million galaxy lenses at over 31 million background source galaxies.

Even the first weak-lensing observations in 1996 were able to confirm that galaxies are much larger than their luminous components. The visible portion of a typical galaxy extends out from its center to a distance on the order of 50,000 light-years. Weak-lensing studies have found that the dark halo continues far beyond this, to at least 400,000 light-years. The edges of dark halos are hard to define, in large part because at a certain point (depending on the environment in which the galaxy is located) the halo from one galaxy begins to overlap with that of a neighboring galaxy, or, if the galaxy is part of a cluster, the galaxy halo begins to merge with the dark matter halo of the cluster.

The typical density profile of a galaxy halo is also starting to emerge from weak-lensing data. Cold dark matter halos have a characteristic density profile: high density in the center, low density at the outer edges. So far, the observed profile is consistent with this generic description, but we haven't yet been able to make out the details that might help us to distinguish between different models.[5]

More recently, astronomers have been using weak lensing to determine the shapes of galaxy halos—are they spherical (like a basketball), or are they somewhat flattened (more like a football)? SDSS astronomers have attempted to use their large database of galaxies to tease out dark halo ellipticities, but the results are inconclusive. There are hints of asphericity, and future weak-lensing studies—with a few more million galaxies—should be able to reveal the shapes of these dark halos.

It's not easy to mine galaxy lenses for information on cold dark matter, mainly because (cosmologically speaking) galaxy halos are too small. The smaller the cosmological object, the higher the probability that its shape has been affected by astrophysical processes such as gas dynamics (which are interesting for understanding how stars form, but a nuisance in attempts to extract the behavior of dark matter and gravity). Smaller lenses also cover smaller areas of the sky, which means fewer sources behind the lens. The innermost regions of galaxy halos are smaller still, making them difficult to observe via any method, including gravitational lensing, so differentiating between cusps and cores in galaxy halos is much more challenging than in larger cluster halos. Clusters remain the premier cosmic dark matter laboratory.

DARK MATTER HALOS, PART II: CLUSTERS

The dark matter halo of a cluster covers an enormous patch of the sky, large enough to contain hundreds or thousands of distant galaxies. Therefore, the dark matter distribution of an individual cluster can be

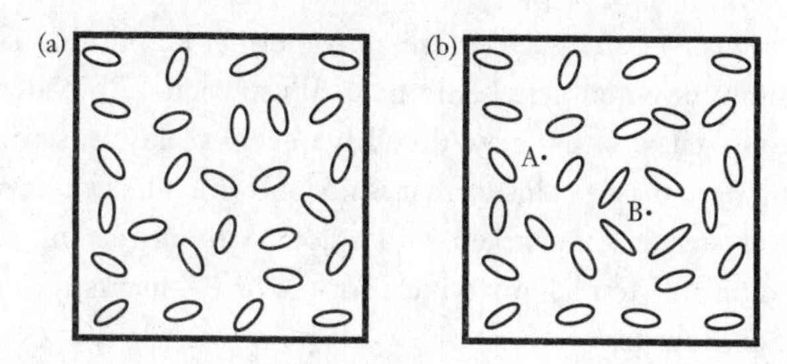

ILLUSTRATION 9.2 The effects of weak lensing. (a) No lensing. All of these cartoon galaxies have the same shape, but point in random directions. (b) Lensing. An invisible lens has been placed in front of the galaxies of panel (a) at either point A or point B. The lensing results in a reorientation of the galaxies about the lens, revealing its location (see text for answer).

reconstructed via gravitational lensing. No stacking or averaging over many clusters is needed. The distortion induced in any one of the galaxies behind the cluster may be tiny, but there are more than enough of these galaxies to reveal the bulk of the dark matter halo. And as we saw in Chapter 7, sometimes a galaxy behind a cluster lens is grossly distorted or multiplied via strong lensing.

The most effective way to map the mass of a cluster is to combine data from both strong and weak gravitational lensing measurements. Strong lensing occurs closer to the center of the cluster, where the mass concentration is the highest. Weak lensing is more efficient farther out—at larger radii the lens covers more background galaxies. Together they paint a fairly complete picture, revealing the dark matter profile of the cluster from its inner regions to its outer edges.

Gravitational lensing is also the best means of probing the number and size of subhalos (the smaller clumps of matter expected in cold dark matter models) within the larger cluster halo. Lensing is blind to the luminosity of the subhalos—they can be bright (lit up by stars within them), faint, or completely dark, but their lensing properties will be the same. Subhalos add to the complexity of the cluster lens, and the resulting distortions of background galaxies—from both weak and strong lensing regimes—will reveal the presence of the subhalos.

Weak lensing by a cluster produces the general effects shown in Illustration 9.2a represents a background field of distant galaxies in the case of no lensing—the galaxies are all somewhat elliptical in shape, and randomly oriented across the field. Illustration 9.2b shows the same cartoon galaxies, but now they have been weakly lensed by an intervening dark mass (a cluster that is not shown). The center of the (invisible) cluster lens is marked with a dot. Without reading ahead, you should be able to pick out which dot—A or B—marks the center of the invisible cluster.

As before, the effects are exaggerated in order to illustrate the idea, so that even without the cluster in the picture you should easily be able to identify the cluster center. All of the galaxies are now slightly tilted in such a way that on the whole they appear to circle about B, pointing the collective finger to give away the location of the cluster lens. In practice, this effect is extracted by a careful statistical analysis of many more background galaxies, so that the signal is unmistakable (the statistical significance is very high). It reveals not only the center of the lens but, when combined with strong lensing, the general topography of the cluster matter.

What have we learned so far from cluster lensing? The answer to this question has two parts. First, clusters allow us to probe the nature of dark matter—exactly how the dark matter is arranged in the halo reflects the behavior of the dark matter particle and so, by studying the shape, density profile, and subhalo population of the cluster halo we can test various particle physics models of dark matter. Second, clusters allow us to use gravitational lensing to test alternative models of gravity. By comparing the location of the lensing mass with the location of the visible mass, we can determine whether a new theory of gravity can eliminate the need for dark matter altogether.

The giant arcs and radial arcs that have been observed in cluster lensing are unmistakable signatures of lensing, but their implications for the central part of the cluster dark matter halo are not yet unambiguous. Giant arcs require a high central density, but the complex nature of a cluster lens (the fact that the lens may be composed of more than one large clump of dark matter—Abell 2218 [see Illustration 7.4 on page 155] is a striking example of this) can also increase the number of giant arcs that are created. The core radius inferred from clusters is smaller than suggested by earlier X-ray measurements, and thus less at odds with cold dark matter predictions, but discrepancies still remain. We don't yet know whether this inconsistency will be settled by more precise computer simulations, more lensing observations—or a new model of dark matter.

Radial arcs offer another clue. A spherical lens can produce radial arcs only if the lens has a core. The observation of a radial arc in several cluster lenses led to the conclusion that cluster halos must have a core, not a cusp—until further theoretical work showed that radial arcs can also be produced if the central part of the cluster lens is not spherical.

A recent high-resolution reconstruction of the mass distribution within five cluster lenses employed both strong and weak lensing techniques and was able to identify galaxy-sized halos within the larger cluster halos.[6] The results were generally consistent with a Universe filled with dark energy and cold dark matter, but more data are needed to really test this model in detail.

The bottom line is that we can't yet draw strong conclusions about the nature of dark matter from cluster lensing, but the field is still relatively young, both theoretically and observationally, and lensing remains one of the most promising arenas for studying the nature of dark matter.

Einstein's theory of gravity, on the other hand, is alive and well. Clusters have not yet told us what dark matter is but, as described below, they have recently confirmed that dark matter exists—and ruled out the possibility that changes to Einstein's theory of relativity can explain the missing mass in clusters.

COLLISION COURSE

Particle physicists love to crash things together. In a tiny fraction of a second, two protons collide and produce a shower of new particles that stream into waiting detectors. Things happen a bit more slowly in the cosmos, but the collisions are just as spectacular—and the debris is just as revealing. The claim of "direct empirical proof of the existence of dark matter" was made in August 21, 2006, by a group of astronomers who found two clusters of galaxies in the aftermath of a major smashup.[7]

The Bullet cluster (which also goes by the professional name of 1E 0657-56) is actually two clusters, and they've been caught in the act. The collision is shown in Color Illustration 8. Seen in visible light, there appear to be two groups of galaxies—one to the left of center in the photo, the other, smaller assembly up and to the right. An X-ray snapshot of the hot gas in the clusters (shown as pink clouds in the image) was captured by the Chandra satellite and reveals the origin of the cluster's nickname. The gas associated with the larger cluster forms an irregular cloud almost in the center of the image, and the hot gas of the smaller cluster has the very distinctive shape of a bow wave—similar to the bow wave that precedes a speeding boat. The gas cloud of the smaller cluster looks like a bullet that has just been shot through the center of the larger cluster.

The apparent bow wave is a shock front that was created when the gas of the smaller cluster plowed through the gas of the larger cluster. Just as the speed of a boat can be estimated from the size of its bow wave (the faster it moves, the larger the wave), the relative speed of the two clusters can also be determined from the bow shock seen in Color Illustration 8—the smaller cluster is moving away from its larger companion at 2,700 miles per second. The reconstruction of the crash reveals that the cores of the two clusters collided about 100 million years ago, and although the centers of the galaxy populations of each cluster are now separated by 2.2 million light-years, the gas clouds associated with each cluster are much closer.

What happened?

A cluster has three main components—the visible galaxies, the X-ray–emitting gas, and, if gravity follows Einstein's (or even Newton's) dictates, dark matter. The bulk of the normal matter is in the hot gas—the gas mass in the Bullet cluster is 10 times higher than the mass in the galaxies. In a cluster that has been quietly minding its own business for a billion years or so, all three of these components will be sitting on top of one another, happily nestled into the gravitational potential well created by their combined mass.

A collision of two comparably sized clusters throws this peaceful cohabitation out of whack, and the various components are, at least temporarily, segregated from one another. During the collision, the gas of one cluster runs smack into the gas of the other cluster, and ram pressure—similar to the drag you feel when you try to move your hand underwater in a swimming pool—slows the forward movement of the two clouds of gas. The galaxies, on the other hand, are for the most part oblivious to what's going on. They're too few and far between to collide with one another, and they don't experience the level of ram pressure that reduces the speed of the hot gas. The galaxies go on ahead, while the gas lags behind.

What about the dark matter? Cold dark matter is as insensible to the gas as are the galaxies—and it doesn't interact with itself in this context either. It will continue moving forward with the galaxies. In a standard cold dark matter Universe, the dark matter and the galaxies will be together, and separated from the gas component of the cluster. This scenario—the segregation of the galaxies and dark matter in one place and the gas in another—is exactly what is needed to differentiate between dark matter and modified gravity. Gravity, even in the alternative gravity models that we examined in the previous chapter, always traces the mass. If there is no dark matter, the vast bulk of the cluster is in the hot gas, so we should find the mass where we see the gas.[8] Alternatively, if dark matter exists, its mass will dominate the cluster and we should find the mass of the cluster where we expect the dark matter to be—centered on the cluster galaxies.

We know where the galaxies are and we know where the gas is—all that's left is to determine where the mass is. A mass map of the Bullet cluster will settle the debate. If the mass coincides with the gas distribution, standard cold dark matter is out. If it lies atop the galaxies, dark matter wins the day.

Marusa Bradac, Douglas Clowe, and their collaborators[9] set out to make such a map with the only cartographic method available: gravitational lensing. The Hubble Space Telescope was instrumental in collecting the data—its pristine view of the sky increased the number of faint galaxies that could be used for weak-lensing measurements, and revealed new strongly lensed systems near the centers of the two clusters. The astronomers combined 10 sources of strong lensing (confirming 6 previously detected arc systems and identifying 4 new sources) with weak-lensing measurements of about 1,900 faint galaxies behind the clusters.

The reconstruction of the mass profiles of each cluster from this data is highly detailed, providing the strong evidence needed to support strong conclusions. The mass of the larger cluster is 2.8 trillion solar masses; that of the smaller cluster is 2.3 trillion solar masses (with uncertainties of less than 10%).

More important, the mass in each cluster lies squarely over the galaxies—not over the gas. Color Illustration 8 sums up the observations, overlaying the images of the galaxies, the gas (in red), and the dark matter (which has been painted on in blue). Dark matter exists.

The strength of this conclusion depends (as do all scientific results) on the statistical certainty of the claim—the accuracy with which the center of each mass distribution can be pinpointed. For the larger cluster, the uncertainty in the location of the mass peak is 10 times smaller than the distance between this peak and the center of the cluster's X-ray gas cloud. For the smaller cluster the mass peak can be located with a precision that is six times smaller than the distance between the mass and the gas. The evidence is clear, and the conclusion of the authors (as reflected in the title of their paper) is strong. The vast bulk of the mass in these clusters is in the form of dark matter.

COLD DARK MATTER

We now know that dark matter of some form exists. This is an incredible statement, and one that could not have been made even a few years ago. Gravitational lensing has played a major role in achieving this state of affairs, but it's important to remember that the evidence for dark matter comes from a variety of independent observations of the Universe on different scales and at different points in its history—many independent witnesses who all give the same description of the perpetrator. We have a very good handle on how much dark matter weighs down the Universe and, in general, where it is. We can't rule out new theories of gravity (new observations or experiments might eventually expand our understanding of gravity), but we can state with confidence that whatever these new theories and their implications might be, dark matter is still a necessary ingredient in the cosmic recipe.

Gravitational lensing studies have confirmed the urgency of direct searches for dark matter particles, and current and future lensing experiments will help narrow the focus of these searches. Further analysis of the Bullet cluster is under way to explore potential constraints on the self-interacting nature of dark matter (the more it interacts, the more it will also lag behind the galaxies), and other cluster–cluster collisions are being studied. More detailed gravitational lensing studies of galaxies and clusters will refine our picture of dark matter halos and either strengthen the agreement with cold dark matter models—or demolish it.

We don't know what dark matter is, but we know that it's out there, dominating the matter sector of the cosmos. Gravitational lensing has outlined the halos of dark matter that enshroud galaxies and clusters, firmly establishing its credentials as a dark matter telescope. We can now refocus this telescope on the truly dark web of matter that pervades the Universe—the dark strands and filaments of matter devoid of any luminous galactic or stellar markers. Imprinted on this web is the influence of an even greater cosmic mystery: dark energy.

CHAPTER 10

An Accelerating Universe

Dark energy is the most compelling mystery in physics today. Observations of distant exploding stars have blown apart our tidy picture of an orderly Universe filled with matter, insisting that we reexamine our understanding of the very nature of space, time, matter, and energy. What strange phenomenon is fueling the accelerating Universe? What do we know about dark energy, this enigmatic and unexpected spacetime propellant? And what does the existence of dark energy imply about the natural world on the most fundamental level? Uncovering the evidence for dark energy is like finding an elephant on top of a table impeccably set with the finest china and silver—adding the napkin rings no longer seems so important. We stare in shock at the uninvited guest and demand to know where the elephant came from—and how it got into a room with no doorway big enough to admit it.

The first place to look for answers to the riddle of dark energy is in the vacuum of "empty" space. We know from quantum physics that empty space is not really empty, but a bubbling cauldron of virtual particles that continuously pop in and out of existence. This lively state of affairs implies that there could be an energy of the vacuum—that the zero point, the lowest energy state, of the vacuum is not really zero. But when we try to calculate the size of this vacuum energy on the basis of our understanding of the subatomic particle world, we find a number that is far too large (by many, many orders of magnitude).

This is a problem, since vacuum energy is also the last place to look within the borders of our Standard Model of the subatomic world. No other explanations for dark energy use physics that is already known and understood. The evidence for dark energy impels us to go beyond our current model of the microcosmos in search of some new physics that will give us a vacuum energy of the right magnitude, or explain the dark energy with something even stranger. Either way, truly understanding dark energy will revolutionize our understanding of the Universe.

THE CASE FOR DARK ENERGY

Physicists are skeptical beasts by nature, and the announcement of data that purport to overturn our basic picture of any area of physics is met with a healthy dose of suspicion. Experimentalists whose results imply a drastic departure from the accepted model of how things work are subjected to an intense interrogation with regard to their methods and analyses, and confirmation in the form of independent data from other groups is demanded. The discovery of dark energy is no exception, especially since its earliest incarnation—the cosmological constant—has been retrieved from the cosmic dustbin on more than one occasion since Einstein chucked it out after hearing of Hubble's data. Each time—until now—it has eventually been tossed back out.

Theoretical rumblings about the need to reintroduce the cosmological constant began to be heard more often in the 1980s and 1990s as measurements of the matter density in the Universe became more precise. Theorists wanted a flat Universe for consistency with their favorite mechanism to explain why the Universe appears to be so homogeneous (a theory known as inflation, which will be discussed more in Chapter 12). A flat Universe requires a critical density of matter and energy, and the theorists expected their colleagues in astronomy to go out and find this requisite amount of matter. The astronomers came up short—the matter found in clusters and galaxies adds up to only about 30% of the critical density, too low to be consistent with a flat geometry. This finding prompted some theorists to suggest resurrecting the cosmological constant, which is a form of energy.

With a judicious choice of a value for the cosmological constant, the total density of matter and energy will be equal to the critical level needed for a flat Universe.

Prior to 1998, however, a flat Universe was a theoretical prejudice, not an observational phenomenon. The possibility that we live in a low-density Universe (which would have a curved geometry) remained a viable option, although one that many cosmologists were not happy with. Cranky theorists cannot take the place of actual data, however, and a healthy skepticism remained. In the absence of a clear signal of dark energy, both models—a flat Universe with a cosmological constant and a curved Universe without one—remained in the running.

As the twentieth century came to a close, this situation changed dramatically. Three independent kinds of observations of the Universe (with several groups working independently on each kind of observation) now provide compelling evidence for a flat Universe whose major component is some form of dark energy.

Observations of the cosmic microwave background reveal the overall geometry of the Universe and provide very convincing evidence that the Universe is indeed flat. Cluster mass measurements find less than 30% of the total amount of matter and energy needed for a flat Universe, making it clear that most of the Universe is in something very different from dark or normal matter. And data from distant stellar explosions known as supernovae indicate that the Universe is accelerating—the signature of dark energy.

EXPLODING STARS

In the first few months of 1998, two groups of astronomers presented their measurements of distant exploding stars.[1] The Supernova Cosmology Project had data on 40 supernovae and the High-z Supernova Search Team reported on 14 supernovae, all at distances greater than 2 billion light-years away.[2] Both teams were composed of experienced observers, who wanted to make sure they had accounted for all the potential uncertainties in their data before making a claim that they knew would rock the scientific community. Saul Perlmutter, head of

ILLUSTRATION 10.1 An exploding star. The bright object in the lower left-hand corner is a stellar explosion known as a supernova. Located about 50 million light-years away in galaxy NGC 4526, a member of the Virgo cluster of galaxies, the supernova was detected in March 1994 and shone brightly for over a month.

the Supernova Cosmology Project, displayed a plot (stamped "Preliminary Analysis") at the 191st meeting of the American Astronomical Society, held in Washington DC in January of 1998, showing that their data implied the existence of a cosmological constant (although he was careful to emphasize the preliminary nature and remaining uncertainties of their data analysis). The High-z team went public the following month at the Dark Matter Meeting at UCLA. The two teams were in complete agreement: these supernovae were fainter than expected for a Universe filled with matter—and some form of dark energy was the most likely explanation.

There are two types of supernovae. The first kind are produced by the flamboyant finale of a star at least eight times more massive than the Sun—a single star going out in style. The second kind—a.k.a. Type Ia supernovae—are the end product of a doomed stellar twosome of low-mass stars in orbit about one another. When one of the stellar duo exhausts its fuel, it becomes a white dwarf star, a cooling ember of carbon and oxygen that, left to its own devices, will simply fade quietly away. The original pair of stars is now a star–white dwarf combo. The trouble begins when the second star begins to follow its own exit strategy. As it runs out of hydrogen fuel, a star first expands in size, puffing itself out into a giant or red giant star (eventually sloughing off its outer layers while the core contracts to form a white dwarf). As the second star goes through the giant stage, it deposits some of its mass on the white dwarf. Not a good idea. White dwarfs are touchy about their weight, and the added mass increases the pressure on the core of the white dwarf, triggering a runaway process of nuclear fusion. The white dwarf explodes in a spectacular fashion, obliterating both itself and its companion.

From a cosmologist's point of view, these explosions are extremely useful. First, they're bright and can be seen at great (i.e., cosmological) distances. Second, they come with a kind of astronomical stamp that certifies their intrinsic brightness, making them excellent distance markers or *standard candles*—or, as some have called them, "standard bombs."

Detailed and copious notes on nearby Type Ia supernovae have revealed that the intrinsic luminosity of these explosions can be determined from how quickly they decline in brightness: brighter supernovae fade away more slowly. Furthermore, the more distant supernovae look just like their closer counterparts. Thus, we can use nearby supernovae, whose distances we can measure by observing other objects in their home galaxies, to calibrate the intrinsic brightness of distant supernovae. It's as if we have learned how to read the fine print on the top of a lightbulb, which indicates whether it's 60, 75, or 100 watts, from across the room—or across the country. Because the apparent luminosity of an object diminishes with distance in a well-known way,[3] a comparison of the apparent brightness of a supernova with its true brightness will yield its distance.

Standard candles at great distances are essential for measuring the expansion rate of the Universe. In an expanding Universe, the recession velocity of distant objects depends on how far away they are—the more distant the object, the faster it is carried off by the expansion of space. Thus, the expansion rate is determined by measurement of the recession velocity as a function of distance.[4] The velocities of receding objects are relatively easily measured via the redshift in their light spectrum (see Chapter 7); determining their distances is much harder.

Edwin Hubble used variable stars as distance markers for the galaxies in his original measurements of an expanding Universe.[5] He looked for stars that follow a regular pattern of variability—growing brighter and dimmer in a consistent and predictable manner. Such stars, called *Cepheids*, had been observed nearby in our own Galaxy, and astronomers found that the timing of their variability was a good predictor of their intrinsic brightness. Hubble homed in on 24 distant stars of this kind in other galaxies. He estimated the distance to each star from the time variations in its light. (The timing implied an intrinsic brightness; comparing this with the observed brightness yields the distance.) Hubble then plotted this distance versus redshift for each galaxy, which had been measured earlier by other astronomers, including Vesto Slipher and Milton Humason. The relationship between the redshift and the distance was striking: more distant stars (and their home galaxies) were indeed receding at a faster rate than those that were closer. Those twice as far away were moving twice as fast. Unfortunately, at that time the calibration of distant Cepheid stars was incorrect (for example, not until the 1950s was it known that there are two types of Cepheid stars, and the relationship between the period of variability and brightness is different for each type), so Hubble's distance measurements were off. As a result, the expansion rate derived from his data is seven times too high.

As distance measurements became more accurate, the estimates of the expansion rate settled down to the current value. The Hubble Space Telescope Key Project, designed to measure the current expansion rate of the Universe, published its final results in 2001.[6] Headed by Wendy Freedman, the Hubble Space Telescope team used various standard candles to probe the nearby Universe (within about 1.3 billion light-years) and found that space is stretching to the tune of 72 kilometers per second per megaparsec, with an uncertainty of about 10%. A galaxy at a distance of 100 million light-years is receding from

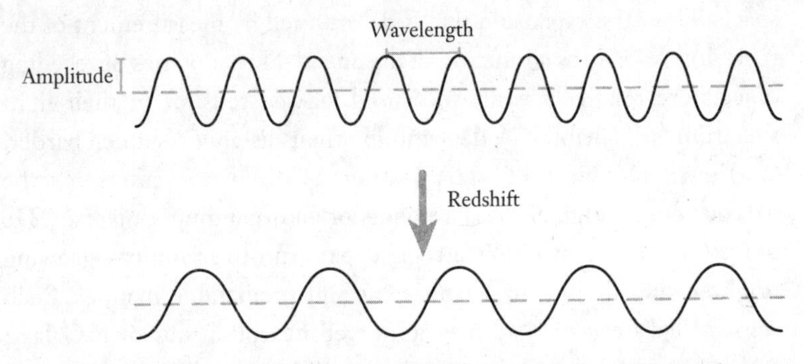

ILLUSTRATION 10.2 A wave. The amplitude (or height) and wavelength (distance between two successive crests or troughs) of an oscillating wave pattern. Redshifting essentially stretches a wave, increasing the wavelength of light.

us with a velocity of about 1,372 miles per second. A galaxy at 200 million light-years is moving away at twice that speed.

The two supernova teams wanted to measure the change in this expansion rate at much earlier times, so they set their sights on targets farther out into the cosmos (and thus farther back in time).[7] On average, the supernovae they found were dimmer than expected, which means that they are farther away than is predicted for a Universe without dark energy—but just where we would expect them to be for a Universe filled with this strange substance.

In an accelerating Universe, distances to remote objects are larger than they would be if the rate of expansion were slowing down. An object with a given recession velocity (as measured by the redshift in its light) in an accelerating Universe is farther away from us than an object with the same recession velocity in a decelerating Universe.

To picture this we need to revisit the redshift discussion. In Chapter 7 redshift was described as resulting from the motion of a light source in the direction away from us. In an expanding Universe there is another way to think about redshift that may help us understand the observable effects of an accelerating expansion of space. Light travels through space as a wave—in this case, oscillations of the electromagnetic field. Waves are described by their amplitude and wavelength. The wavelength is a measure of the distance between the crests of the wave; the usual picture of a wave is shown in Illustration 10.2. As a light wave travels through an expanding Universe, its wavelength is stretched along with the stretching of space. Think back to the rubber sheet analogy for space, and draw a "light wave" on the sheet. Now stretch the sheet—the wave will be stretched as well, and its wavelength will increase. The light will be redshifted. The more the sheet is stretched (i.e., the more space has expanded), the larger will be the redshift in the light.

The redshift of the light from a distant source—a supernova, galaxy, or quasar—is a measure of how much the Universe has expanded since the light was emitted. The same amount of expansion takes a longer time in an accelerating Universe than in one that is slowing down. If two cars pass you at the same speed, but one has been accelerating while the other has been braking since they passed the last checkpoint, the one that is speeding up took longer to travel the distance from the checkpoint to you—it was moving more slowly before it got to you than the braking car was. A longer expansion time means that more time has elapsed between when the light was emitted and when we detect it. More time means that the light, zipping along at the cosmic speed limit, has traveled a longer distance—and thus the source of the light is farther away, and will appear dimmer to us.

This effect was observed by the astronomers looking at high-redshift supernovae. These stellar explosions were fainter than expected for a Universe whose main component is matter, and the inferred distances are larger, implying that the expansion of space must have been slower in the past. The Universe is accelerating.

More recent supernova data are giving us an even more detailed picture of the cosmic expansion history. The Universe was not always accelerating. The fact that the Universe contains galaxies means that even if the Universe today is dominated by dark energy, there must have been an earlier epoch when matter was the main component of the cosmos. Galaxies can form only in a matter-dominated Universe. By looking further back in time, at even more distant supernovae, we should be able to find evidence of this matter-dominated era in the form of a deceleration of the expansion rate.

Astronomers have recently documented the transition region where the Universe changed gears. In November of 2006 Adam Riess, one of the leaders of the original High-z Supernova Search Team, and his collaborators reported the detection of 23 supernovae at distances much greater than 7 billion light-years. These data confirm the picture of a recent period of acceleration, due to the dominant influence of dark energy; and an earlier era of deceleration, when matter had the upper hand. As it expanded from the initial hot Big Bang, the Universe first reduced its speed as the mutual gravitational attraction of matter and radiation energy put on the cosmic brakes; then, roughly 5 billion years ago, dark energy hit the accelerator and the Universe took off.

These new results also put an end to another explanation for the dimmer-than-expected supernovae. How do we know that there isn't some kind of dust spread throughout the Universe that is muting the apparent brightness of these distant exploding stars? Dust can obscure or dim the light from astronomical sources, but in general the absorption and reemission of the light changes the color of the original source. The distant supernovae appear to have the same colors as their local counterparts, so it would have to be some kind of heretofore undetected "gray" dust whose only effect was to dim all colors of light equally. However, such a dust would have the biggest effect on the most distant supernovae, suppressing their light output by a larger amount (a larger distance implies that more dust is blocking the light). Instead, the data show that the dimming of the most distant supernovae is less than it is for those slightly closer—exactly what is expected for a Universe that was slowing down before it began to speed up.

The supernova data provide convincing evidence of the acceleration of the Universe—and the presence of some kind of dark energy. However, these data alone cannot specify the amount of dark energy. Supernova constraints on dark energy depend on the amount of matter in the Universe. The more matter there is in the Universe, the more dark energy is needed to overcome the gravitational deceleration due to this matter and to boost the expansion to the observed accelerated rate. We need more information to determine the amounts of dark matter and dark energy in the cosmos. Fortunately, the cosmic microwave background data provide the perfect complement to the supernova results.

THE ULTIMATE GEOMETRY LESSON

As mentioned in our bug world explorations in Chapter 3, triangles obey different rules in different geometries. Shown in Illustration 10.3 are three different kinds of two-dimensional surfaces, representing closed, flat, and open space. On each surface a triangle has been drawn, and the side marked "L" is exactly the same length for all three triangles. The distance between this side and the opposite angle (marked Θ) is also identical for all three. What is not the same is the size of this angle—it's larger in a closed universe than a flat universe, and smaller in an open universe.

To measure the geometry of the Universe, we just need to find a suitable triangle. This means finding a standard ruler—an object that

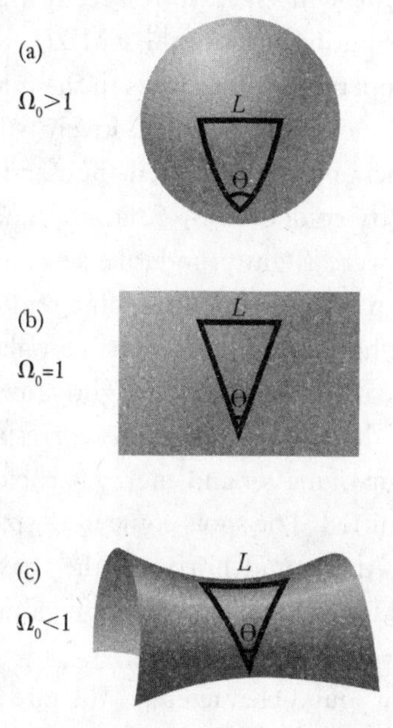

ILLUSTRATION 10.3 Two-dimensional examples of curved geometry. (a) Positively curved space corresponding to a closed, or overdense, universe. (b) Flat (zero curvature) universe with the critical density. (c) Negatively curved space corresponding to an open, or underdense, universe.

has a known length *L*—that sits at a large distance away from us (the geometry-induced difference in the observed angle is larger for larger distances) to act as the far side of the triangle. The apparent size of this ruler—the angle it subtends on the sky—will reveal the overall curvature of space. This is where the cosmic microwave background (CMB) comes in. Embedded within the map of the microwave sky is the perfect standard ruler.

The average temperature of the microwave background is 2.725 degrees above absolute zero, and this temperature is almost exactly the same all across the sky—but not quite. It varies at the level of plus or minus about 0.0002 degrees Kelvin. George Smoot won the 2006 Nobel Prize in Physics for the first detection of these minuscule temperature variations by the Cosmic Background Explorer (COBE) satellite in 1992. His colleague John Mather shared the prize for COBE's 1990 measurement of the CMB spectrum, which revealed the temperature of the Universe. More recent measurements have produced higher-resolution maps of the CMB.[8]

The tiny temperature differences in the microwave sky produce a pattern of hot and cold spots (relatively speaking—negative 455 degrees Fahrenheit, give or take a ten-thousandth of a degree here and there, is still pretty cold, even by Chicago standards) that correspond to regions that were slightly underdense or overdense in the early Universe. Light particles—photons—in a region of higher mass density are a tiny bit hotter than the average. Similarly, light that starts off in an underdense region will have a slightly lower temperature.

The sizes of the hot and cold spots correspond to the sizes of the ripples in the soup of matter and energy particles at the time the CMB photons were emitted. The spots come in all sizes, but the most typical size is given by the sound horizon—the maximum distance that a sound wave could travel through the cosmic fluid at the time the CMB was generated.

The speed of sound characterizes the rate at which a disturbance of some kind can propagate through a particular medium. For example, the sounds we hear are the result of the motion of air molecules. The vibration of a drumhead causes the air molecules next to the drum to oscillate; these air molecules jostle the ones next to them, which then bump into their neighbors, and so on until the wave of oscillating molecules hits our eardrum and causes it to vibrate. In the early Universe, the cosmic fluid is jostled by the collapse of matter onto small overdensities, creating waves that propagate through this fluid. The speed of sound (which is almost 58% of the speed of light at this time) limits how far one of these waves can travel.

The waves are composed of normal matter and light, which are tightly coupled to one another in a hot plasma. These waves oscillate about minute clumps of dark matter—regions where the density is just a tiny bit higher than average (100,006 particles over here compared to 100,000 particles over there), and also about regions with lower-than-average density. Gravity and pressure once again work in opposition: gravity pulls the plasma in toward the lump of dark matter; pressure pushes it back outward again when the photons object to the compression. The two continue their tug-of-war over the plasma, and a pattern of oscillating waves is set up. The particle soup looks a bit like the surface of a pond that has been disturbed by a shower of gravel falling into it, with a ring of waves propagating outward from wherever a stone hits the water.

When the temperature of the Universe drops low enough for neutral atoms to form, the photons decouple from the normal matter and the oscillations come to an end. The normal matter can now fall into the dark matter clumps, unimpeded by the reluctance of the photons. And the photons, no longer dragged in toward the dark matter overdensities by the normal matter, are free to stream off into the Universe. These particles of light carry with them the imprint of the ripples in the cosmic fluid that existed less than 400,000 years after the Big Bang.

No one was around at that time to listen to this ringing of the early Universe, but we can detect it today through our CMB telescopes. The ringing has a fundamental tone, just like a bell, whose pitch (wavelength) is given by the sound horizon at the time the CMB is created—the largest distance a wave could have traveled through the hot plasma. The fundamental tone depends on the speed of sound in the

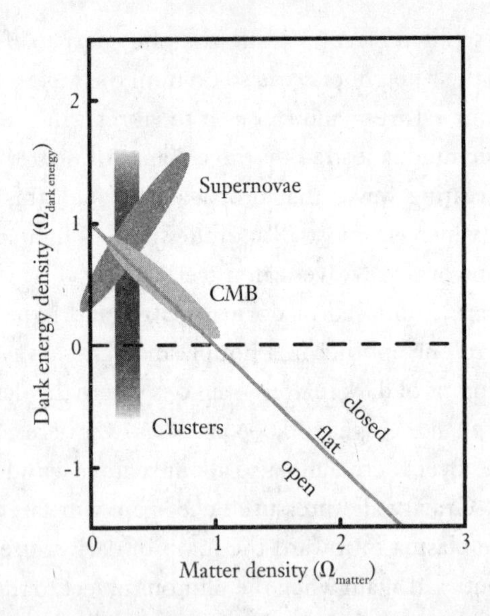

ILLUSTRATION 10.4 Measurements of the cosmic microwave background (CMB), the matter density in clusters, and the distance to exploding stars (supernovae) independently place constraints on various combinations of the amount of dark matter and dark energy in the Universe. The size of the gray areas is meant to indicate (roughly) the uncertainties in each type of measurement—these areas continue to shrink in size as more, and more precise, measurements are being made.

plasma (which is fairly well-known) and the amount of time between the Big Bang and the release of the CMB. This tone is imprinted on the temperature map of the sky as a hot or cold patch, and thus the intrinsic size of a typical spot can be used as a standard ruler. The apparent size of a typical spot as seen in the maps of the microwave sky can then be used to determine the geometry of the Universe. In a flat Universe, the most common spot size is about 1 degree across (for comparison the Moon is about half of 1 degree across); in an open Universe the typical spot size will be smaller, and in a closed Universe it will be larger.

On April 27, 2000, the Boomerang collaboration published the first map of the CMB with enough detail to measure the size of the hot and cold spots.[9] (There were earlier hints from the Microwave Anisotropy Telescope [MAT] located in the Chilean Andes, and another balloon-borne experiment, MAXIMA, published its data 2 weeks after Boomerang). Boomerang found that the typical hot or cold patch on the microwave sky is 1 degree across—the Universe is flat.

The combination of the results from microwave background experiments, distant supernovae, and clusters is shown in Illustration 10.4.[10] We live in a flat, accelerating Universe, whose matter density falls far short of the total. Most of the Universe is in the form of dark energy. Now we just need to figure out what dark energy is.

THEORIES AND WILD IDEAS

The playing field is wide open. Theorists are taking full advantage of the current window of opportunity to run wild through the vacuum of empty space, inventing new substances and fields and dabbling their toes in extra dimensions. As with the dark matter conundrum, the current conflict between theory and observations can be resolved by the addition of something new to either the right-hand side of Einstein's equations—a new kind of energy; or the left-hand side—a new formulation of gravity.

Setting aside modifications of gravity for the moment, there are essentially three categories of dark energy proposals:

1. *Cosmological constant.* Vacuum energy (energy of empty space) that is constant throughout space and over time.

2. *Quintessence.* A dynamic version of dark energy—its value changes over the course of cosmic history.

3. *Other.* A heading that encompasses ideas a bit further off the beaten path (including theories still to be invented).

The one common factor among all three of these options is negative pressure. In the Universe the pressure of a substance adds to (or, in the case of negative pressure, detracts from) the braking power of gravity.

In general relativity, all energy couples to gravity—including the kinetic energy of a hot gas (its pressure) and the radiation pressure of light. These pressures add to the gravitational self-attraction of the gas or the light. The strength of gravity depends on the both the mass and the pressure of the gas or light. Thus, it's harder to expand a Universe filled with a substance that has a (positive) pressure than one that has no pressure.

Pressure can be positive or negative or zero. The pressure of gas or light is positive, but in theory, a substance could exist that has a negative pressure. Negative pressure, in Einstein's equations, has the opposite effect on the strength of gravity: it induces a measure of gravitational repulsion and decreases the total gravitational self-attraction of the cosmic fluid. If the pressure is negative enough, such a substance can override the deceleration of the Universe due to the matter within it, and the Universe will accelerate.

Matter (normal or dark) has zero pressure. Imagine a collection of (cold) matter particles spread across a volume of space. Their energy density is simply the energy of their mass, and the gravitational self-attraction of this mass energy opposes the expansion of the Universe. In a pseudoeconomic version of the cosmic equations, the Universe has to pay a tax based on the energy density in the mass. This "tax" slows down the rate of expansion.

Radiation, in the form of light particles or fast-moving (relativistic) matter particles, has an associated positive pressure whose value is tied directly to the level of its energy density—a decrease in the density of radiation is accompanied by a decrease in the pressure. The expansion has an extra passenger in the form of the pressure, and the cosmic tax collector is counting heads—the tax that is exacted is based on both the energy density and the pressure of the gas. Thus, radiation slows the expansion even more effectively than pressureless matter does.

Dark energy behaves very differently. It has a negative pressure. Negative pressure decreases the gravitational self-attraction of a substance—the more negative it is, the stronger its self-repulsion. Instead of an extra tariff for the pressure component of dark energy, its negative pressure effectively results in a tax rebate. Overall, this little kick-back scheme encourages the expansion and the Universe accelerates.

Einstein's equations and the conservation of energy spell out the detailed requirements for achieving this acceleration. The expansion will slow down because of gravitational self-attraction if the sum of the total energy density plus three times the pressure is positive; it will accelerate if this sum is negative.

For each kind of matter and energy, the relationship between density and pressure is encapsulated in an effective equation of state, which describes how "springy" or "squishy" the substance is. Physicists characterize the different kinds of matter and energy by the ratio of the pressure p to the density. This ratio is known as w:

$$w = p/\rho$$

For matter, $w = 0$ (pressureless); for radiation $w = +\frac{1}{3}$ (positive pressure); for the traditional cosmological constant $w = -1$ (negative pressure). If you were able to confine each of these substances in a box, the matter would simply sit there; the radiation would push out on the walls of the box; and the dark energy would try to pull the walls in.

The effect of a substance with negative pressure on the expansion of the Universe depends on exactly what the Universe contains—how much dark energy, what kind of dark energy (the value of w), how much matter, and how much radiation. In a Universe with only dark energy, spacetime accelerates if $w < -\frac{1}{3}$. In a Universe like ours, which also contains matter, w must be even more negative—the more matter there is, the more gravitational self-attraction there is to overcome. And in most theories, w must be greater than or equal to -1 to avoid introducing unwelcome consequences such as particles that travel faster than the speed of light.

The different models for dark energy thus aim for theories with w somewhere in the range of -1 to $-\frac{1}{2}$ (and current data seem to favor values near -1), although some have ventured past the lower limit into the regime where w is less than -1. Such a model wins a place in the coveted "other" category.

THE CONTENDERS

Before starting off on a brief tour of some of the dark energy candidates that have been proposed to date, it's worth pointing out again that none have yet emerged as the most likely to succeed. And it may be that none of them will. The best we can do at the moment is to group the various ideas into general categories and then try to narrow down the possibilities by putting them through a series of observational challenges.

Option 1: Cosmological Constant

The cosmological constant (often called lambda, or Λ) is the most familiar to physicists. It's been lurking in the shadows for the past 90 years, and our current understanding of the subatomic world implies that the cosmological constant should exist—unless it is somehow exactly zero. According to the Standard Model of particle physics, there is an energy associated with the vacuum of empty space. Quantum mechanics allows for the creation of virtual pairs of particles, a process in which a particle and its antimatter twin bubble in and out of existence in the briefest flicker of time—and in quantum mechanics, anything that is not forbidden is mandatory. Thus, the vacuum is actually a sea of energy, composed of contributions from every possible assembly of these virtual particles.

Attempts to calculate the value of this energy lead to major problems. The first answer that pops up is infinity. The contributions of the particles and fields in the standard model get larger and larger as the energy of the virtual particle pair increases. There are simply more ways to create these particles at ever-higher energies, all of which must be added to the total. This unruly behavior is not a problem for calculations within the Standard Model itself, which does not include gravity. Without gravity, the absolute value of the vacuum energy is irrelevant—only differences in vacuum energies ever come into the equations. (The infinities are simply lumped together and put aside.) Think of riding a bike somewhere near sea level. For most of us the major concern is how much altitude is gained or lost over the chosen route—how many hills must somehow be painfully pedaled up. When gasping for breath after a steep climb of 500 feet, the issue of whether the hill stretched from 10 feet above sea level to 510, versus starting at 20 and ending at 520, is moot.

Gravity, on the other hand, insists on a number for the vacuum energy. Spacetime responds to the actual value of the vacuum energy, so we need a way to calculate this value that leads to a sensible answer. The preceding calculation, which yielded a vacuum energy density of infinity, assumed that quantum mechanics is valid to arbitrarily high energies, but we know this isn't true—quantum mechanics can take us only so far before it can no longer be trusted. At very high energies, the quantum mechanics of gravity becomes important. We don't know exactly what happens at these energies, but since gravity is linked to the nature of spacetime, it's bound to be something on the far side of strange.

A second approach to calculating the energy of the vacuum makes a different assumption—that above a certain "cutoff" energy, the underlying theory (which we don't yet know) somehow prevents any further additions to the total vacuum energy. We add up all the contributions of all the virtual particle pairs at energies all the way up to the cutoff energy, but no further. The resulting answer is finite. Adding all the numbers between 1 and infinity will always yield infinity; adding all the numbers between 1 and 1 million produces a large but finite answer (which is left as an exercise for the reader).

The most obvious cutoff energy to use is the energy at which we might expect quantum gravity to rear its head, known as the *Planck scale*. The Planck scale is approximately 19 orders of magnitude larger than the mass energy of a proton. (Cosmologists tend to stick with an order-of-magnitude estimate here—a reasonable approach, since the value of the vacuum energy calculated by imposing a Planck-scale cutoff is so enormous that any number out front—2 or 5 or 7, for example—is completely irrelevant.) The energy density calculated in this way leads to a prediction for the dark energy density of about 10^{120}. Because observations point to a much lower number for the dark energy density, about 0.7, this value is off by only 120 orders of magnitude (a 1 followed by 120 zeros). Oops.

A next step is to invoke supersymmetry, a theory that was discussed in Chapter 8. If supersymmetry does exist, it implies a new partner for every known particle. The fact that we haven't yet seen any of the supersymmetric particles means that they must be much heavier than their normal-matter counterparts. (Or that they don't exist, of course, but let's assume for the moment that they do.) If the electron's superpartner, the selectron, has the same mass as the electron, it should already have shown up in particle accelerator experiments. Because the selectron hasn't yet been detected, it must be heavier than the highest energies we've been able to produce in accelerators so far, and thus its mass is larger than the mass of the electron. This means that the symmetry between the electron and the selectron is broken.

We don't yet understand exactly how supersymmetry is broken (there are a lot of ideas, of course, but no definitive mechanism as yet), but symmetry breaking in general is relatively easy to understand.

Consider a roulette wheel. Initially, the wheel and ball are spinning rapidly and all numbers are equally possible (assuming the wheel is fair). There is a symmetry among all of the numbers—they're all equivalent, and none has been singled out as special. But when the wheel and ball slow down, this symmetry is broken. One number will be chosen and only that number will pay off. Similarly, supersymmetry may be part of an underlying theory of the particle world, but it is broken in the real Universe.

The energy at which supersymmetry is broken gives us another chance to limit the infinities. Above this energy the vacuum energy contribution of each Standard Model particle is exactly canceled out by the contribution of its supersymmetric partner. The cutoff scale for contributions to the vacuum energy in this theory is the supersymmetry breaking scale. The lowest estimates for this energy are about 1,000 times the mass energy of a proton, which lowers the expected value of the vacuum energy of the Universe to a mere 10^{60} or so. Still not even close.

This is a very real problem. Even if dark energy is explained by one of the other options listed earlier, and not by a cosmological constant, we still have to explain why the predictions for the vacuum energy value are so far off. With or without supersymmetry, quantum mechanics predicts a very large value for the vacuum energy that is simply not observed. If the vacuum energy were as high as 10^{60}, for example, we wouldn't be here to worry about it—no stars or galaxies or planets could ever form in such a Universe. Thus, we know that something important is missing from our current picture of the subatomic arena, even if we don't yet know what that something is.

The standard resolution to this problem is to assume that some as yet unknown physics sets the value of the vacuum energy to zero—theoretically, it's easier to make something exactly zero than to fine-tune its value to a number just a little bit above zero. The actual mechanism is not known, but the hope is that a unified theory of all the forces will ultimately provide the answer. Possibly some other symmetry in this larger theory will provide a cancellation mechanism that will nullify most or all contributions to the vacuum energy. In any case, the cosmological constant is giving us additional information about the Universe and its laws—hints that might eventually help lead us to the next level of understanding.

Observationally, the hallmark of a cosmological constant is a measured value of $w = -1$ that doesn't vary in time. A substance with this equation of state, where the pressure is equal to the negative of the density ($p = -\rho$) has an energy density and pressure that do not change with the expansion or contraction of the Universe, but remain constant throughout the history of the Universe. A cosmological constant is indeed constant.

This raises another question: why did dark energy become the dominant component of the Universe just now? (Cosmologically speaking, 5 billion years ago counts as now.) The energy density of the cosmological constant was exactly the same in the early Universe as it is today, while matter and radiation had much higher energy densities at earlier times. Thus, in the first moments of cosmic history, radiation (and later matter) outweighed the cosmological constant by an enormous amount (many orders of magnitude). Yet today, matter and dark energy densities are roughly equal—a cosmic coincidence that troubles and intrigues physicists. Sometimes a coincidence is just that—but it might also be a portent of a deeper theory. We don't yet know.

Option 2: Quintessence

Quintessence—the fifth essence—is dynamical dark energy.[11] Its energy density is not constant over time, and may have been larger in the past. Or not. And what it will do in the future is completely up in the air. This is where our newfound uncertainty about the ultimate fate of the Universe arises. If dark energy is a cosmological constant ($w = -1$), we know what awaits us: the Universe will expand forever. Hundreds of billions of years from now, long after the Sun has gone out and the Earth no longer exists, the view from wherever our descendants (if any) are living will be much less interesting. The Milky Way and its neighboring galaxies will become increasingly isolated as distant galaxies are carried out of view by the continued acceleration of the expansion of spacetime, until none are visible but those within the local group.

If dark energy is some form of quintessence, however, anything is possible. It all depends on the evolution of the quintessence, and many models have been proposed. Quintessence has a value of w that is greater than -1 but less than $-\frac{1}{3}$, and an energy density that in general decreases as the Universe expands. Furthermore, the equation of state for quintessence can change over time. The value of w itself can be time-dependent such that the pressure of quintessence becomes either more or less negative as the Universe evolves, which also plays head games with the expansion rate.

Most models of quintessence start with the assumption that the cosmological constant we've discussed is exactly zero. Next, something new is added to the cosmic mix. It might be a new energy field, a new condensate of ultralight particles, or even a tangle of defects in spacetime. Whatever it is, the pressure of this new component is negative, providing the needed cosmic accelerant, and its energy density is smoothly distributed across the Universe, declining, for the most part, to clump in galaxies and clusters like matter.

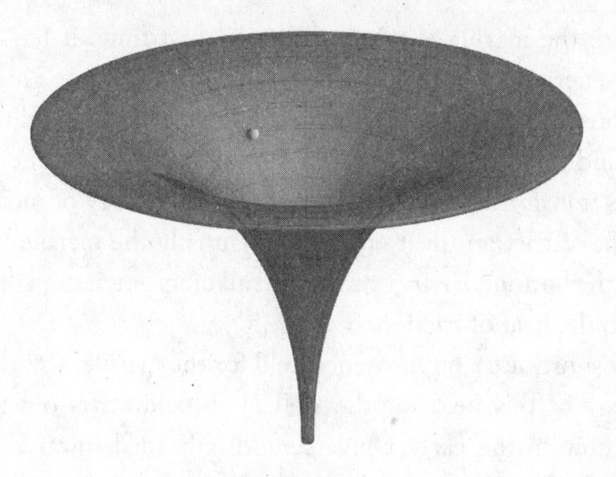

ILLUSTRATION 10.5 A toy model. The behavior of a marble falling into a dimpled depression, or "well," can help illustrate some of the ideas behind quintessence models (see text).

The simplest model for quintessence is a new energy "field" (an energy that is everywhere in space) that is not at its lowest value. For reasons that vary among models, this energy field is decaying toward a minimum level but hasn't yet gotten there. Think of a spring that has been stretched and then placed in a tub of a very viscous fluid, such as molasses. The spring will slowly return to its unstretched state, but until it reaches this completely relaxed position it has a potential energy associated with it. Similarly, until the quintessence energy field reaches its lowest state it has a potential energy.

Fields and energy levels sound more complicated than they actually are, and a simple toy model can help illustrate the basic idea. Picture a marble rolling down into a well with smooth sides as shown in Illustration 10.5. If you release the marble near the top edge of the well, it will roll to the bottom (picking up speed on the way down), then head back up the other side until it runs out of steam and rolls back to the bottom and up once more, and so on. On each subsequent roll back up the side of the well the marble reaches a height slightly lower than its high point on the previous trip. (We're assuming that this is a real marble and a real well, so there is some loss of energy to friction.)

Before the marble is released for the first time, it has potential energy—energy it can potentially extract from gravity. This gravitational potential energy depends on the difference in height between the top and the bottom of the well. As the marble rolls, the potential energy is transformed into kinetic energy (the energy of motion) plus a little bit of friction (heat energy). Eventually the marble will settle down at the bottom, having transferred all of its gravitational potential energy to the heat of friction.

Now substitute a quintessence field for the marble, and the potential energy of this field for the well. The field starts out at a high energy value in the early Universe and is headed toward its lowest energy state (the bottom of the well)—but it hasn't gotten there yet. It's rolling very slowly (in our toy example we could get the marble to roll slowly by making the sides of the well almost flat, with just a small tilt toward the very bottom), so at the present time there is still a lot of potential energy associated with the field. This potential energy acts like a negative pressure, and fuels the acceleration.

As the Universe expands, the field will continue to roll toward zero, decreasing the potential energy and thereby changing the quintessence equation of state—in effect w will become less negative. One possible outcome of such a model is that this potential energy ultimately gets transformed into new matter particles (the heat friction in our toy model corresponds to particle creation), and the Universe will again become matter-dominated and reenter a phase of decelerated expansion. Theorists expect that a mechanism similar to this kind of quintessence propelled the Universe into a period of accelerated expansion, known as inflation, very early in its history. We'll discuss inflation in more detail in Chapter 12.

In most quintessence models the dark energy density decreases with time, but in some it increases. Motivated by the cosmic coincidence problem, there is even a variation of quintessence in which the dark energy density tracks the behavior of the radiation and matter. In these models, at early cosmic times the energy density in the quintessence field latches onto the radiation component and mimics its behavior, acting as a kind of cosmic parasite. When matter energy begins to dominate over radiation, the quintessence jumps off its radiation host (no point staying with a losing team) and switches its allegiance to the matter density. The challenge in these models is finding a mechanism for convincing the quintessence field to stand on its own, so that it locks into the energy density we observe in dark energy today.

Option 3: Other

By definition, this category is incomplete. Scientists continue to generate outlandish new ideas, and the final answer to dark energy may well be a member of this cohort—one that has not yet been dreamed up. Most of these "other" models are less palatable to physicists than those in the first two categories, and in some cases they have been suggested mainly to point out just how far the limits can be pushed. The dramatic consequences of these theories are illustrated by two rather extreme proposals.[12]

The first example is a combination platter with two entrées—a cosmological constant plus quintessence. In most quintessence models the cosmological constant is set to exactly zero and the dark energy is purebred quintessence. But since we don't have the official papers, we have no proof that dark energy might not be a bit of a mutt. Such a model has double the challenges because it introduces two new bits of physics that have yet to be documented, but this feature also opens the door to another kind of ending to the story of our Universe. When the quintessence finally decays away, the energy density of the vacuum will dominate the total energy density of the Universe.

In our earlier discussion of a cosmological constant model (with no quintessence) we assumed that the value of the vacuum energy was positive, but in principle, nothing forbids a small negative value. In practice, it's a disaster because a negative value for the energy of the vacuum—a negative cosmological constant—works in the same direction as gravity and causes the Universe to decelerate even faster, obviously not consistent with what we observe. If we add quintessence to the picture, however, the sum of the energy density in quintessence plus the energy density of the vacuum can be positive even if the bare value of the vacuum energy is negative. The quintessence field will eventually decay and the driving component of the Universe will be the (negative) vacuum energy, which will put on the cosmic brakes and reverse the expansion. A Universe with a negative cosmological constant has only one place to go: it will eventually recollapse in the Big Crunch.

The second example portends an even more gruesome fate for the cosmos. The role of dark energy is played by *phantom energy*—a hypothetical substance with a truly bizarre equation of state. Phantom energy has a negative pressure with $w < -1$, which implies that its energy density will grow as the Universe expands. If we plug phantom energy into Einstein's recipe for the Universe, we find that as the volume of the Universe increases, so does the energy density in phantom energy. A Universe filled with this phantom energy will accelerate at an ever-faster rate—in the cosmic tax analogy, the Universe gets paid to expand. It's the ultimate binge and it doesn't end well. As spacetime stretches at an increasingly frenetic pace, it begins to rip apart clusters, then galaxies, then stars and planets—ultimately shredding apart every atom in what is affectionately known as the Big Rip.

Such a wild ride doesn't come without penalties, of course. Phantom energy can lead to particles with superluminal (faster than light) speeds or a vacuum that completely freaks out. In the construction of new theories, a basic requirement is that the vacuum—which is, in a sense, the ground on which we build the theory—is stable. In phantom energy models, it's possible for the vacuum of "empty" space, with its quiet sea of virtual particles, to quickly decay into total chaos—a situation analogous to attempting to build a house on quicksand during an earthquake. But it's also possible to find theories in which these potential catastrophes can be avoided, so phantom energy can't yet be excluded as a possible solution to the dark energy problem.

MODIFIED GRAVITY REDUX

Which brings us back to gravity itself. Modifications to gravity cannot explain the evidence for dark matter, but this doesn't—and shouldn't—

deter theorists from exploring the possibility that a new variation of general relativity could be responsible for dark energy. It's possible that general relativity has finally come in for an upgrade.

The general goal of alternative gravity models is to weaken gravity at late cosmic times (within the past 5 billion years or so) and over very large scales. Such models are not easy to construct and have to pass at least two kinds of tests—constraints from the early Universe (we still need to be able to form galaxies) and observations of gravity at work in the Solar System, where gravity has been well tested.

Modifications to Einstein's equations can alter the light-bending predictions of general relativity by effectively adding a new force that causes light to deviate from the pure geometric path dictated by the curvature of spacetime. Recent measurements within the Solar System have limited such deviations to extremely small values. The Cassini spacecraft was launched in 1997 and reached the planet Saturn in 2004, spiraling into orbit about this ringed planet in order to fulfill its main mission of obtaining exquisitely detailed images of Saturn and its moons (it has at least 34 moons).[13] During the 7-year journey to the ringed planet, Cassini was also put to work testing general relativity, via a mini-gravitational lensing experiment. In 2002, when Cassini was on the far side of the Sun from the Earth (the Earth, Sun, and Cassini were in nearly perfect alignment), a light signal was sent from Earth to the spacecraft and back again, passing close to the Sun on both legs of the trip so that it would be gravitationally deflected by the warp in spacetime created by the mass of the Sun. The timing of the trip agreed with general-relativistic predictions to within 0.002%, putting tight constraints on any modifications to Einstein's theory.

The orbital path of the Moon about the Earth can also be used to test our theory of gravity. Apollo astronauts (missions 11, 14, and 15) positioned mirrors on the surface of the Moon specifically for lunar laser-ranging experiments.[14] A laser beam is sent from Earth to the Moon, bounced off of the mirrors, and returned to a detector on Earth. The precise timing of the round-trip is used to measure the distance to the Moon to within 1 millimeter, and the orbit of the Moon can thus be traced out in great detail. The results of these measurements are in beautiful agreement with general relativity and also severely curtail any expanded versions of this theory.

Nonetheless, a smorgasbord of modified gravity models has been proposed (and the list continues to grow).[15] One model spices things up by tossing in a new graviton that has a tiny mass. (A *graviton* is a particle that transmits the gravitational force, and is usually assumed to be massless.) A massive graviton might be fine over short distances, obediently carrying out the orders of Newton and Einstein, but over longer distances it's likely to decay into other, lighter particles before it can complete its assigned task. This loss of gravitons (through decay) effectively weakens the force of gravity over large distances. Other models attempt to add new terms to Einstein's equations, which is not as easy as it sounds. The simplest of these models was already rejected, because it was shown to effectively introduce a new long-range force that conflicts with the Solar System tests of gravity. More complicated versions of this type of model have been crafted to minimize the damage.

The real fun begins when we expand our search to include theories with more than four dimensions of space and time. Analogies along the lines of the bug world can help us visualize the basic ideas of these models. First, imagine that our usual (four-dimensional) Universe is an infinite sheet of spacetime similar to the illustrations in Chapter 3. We ourselves, and all of the normal-matter particles, force carriers (except the graviton), and dark matter particles, are confined to live on this sheet, so in a sense it is the entire Universe to us. But unlike our bug world example, there is more. The sheet—called a *brane* (as in *membrane*)—is embedded in a higher-dimensional Universe. Only gravity is allowed to wander off the brane and explore these extra dimensions. The rest of us are confined to the brane.

This confinement can be accomplished in string theory, where particles are not pointlike, but oscillations of tiny bits of string, and each end of the string is anchored to the brane. No loose ends are allowed, so it is forbidden for a bit of string to detach itself and move off into the extra dimensions. The graviton in these theories is also described as a string, but instead of a length of string, a graviton is a tiny loop. Loops have no ends, so the restriction on loose ends doesn't apply to gravitons. They are free to move along the brane or off of it, and this extra degree of freedom effectively weakens the force of gravity. Extra dimensions are analogous to a showerhead with several settings. Turn the handle so that the water shoots out in a single jet (in essentially one dimension) and you'll feel a much stronger force from the water than if you twist the showerhead to a spray setting, which produces a gentler and broader (two-dimensional) shower. The more dimensions that gravity is free to explore, the weaker it appears at any given point in the Universe.

In one model of a brane Universe, the loss of gravitons from the brane to the extra dimensions induces a warp on the brane—a positive curvature that acts very much like a cosmological constant. In another, there is more than one brane in the bigger, higher-dimensional Universe—in effect more than one Universe like ours. In this scenario the Big Bang was triggered by the collision of two branes: ours and another parallel brane. The colliding branes move apart from each other after the crash, but they get only so far before they are pulled back toward each other and another collision, in an unending cycle of crash and separate. Each time the branes head back for another go at each other, the spacetime on the branes begins to accelerate. The supernova data could be a cosmic signal that we're ramping up for yet another brane–brane impact.

Of course, there are challenges in creating such theories. In order to induce the observed behavior of the Universe, brane theories must somehow control the escape of gravitons into the extra dimensions so that they don't conflict with our observations of gravity within the Solar System. And overall, such theories seem to be plagued with various problems, such as nasty little instabilities called *ghosts*. Ghost particles can introduce negative probabilities into the theory, making them extremely unwelcome theoretical additions. Ghosts are a sign that the theory as formulated is not viable—and more work needs to be done.

That more work needs to be done is the general theme of dark energy models. It's obvious that the theories that have been proposed, in however many dimensions their inventors seem to be most comfortable, are wonderfully ambitious and inventive—and they need to be. The explanation for dark energy is likely to be something on the wild side, and new physics is clearly called for. The answer may be similar to one of the preceding examples—or it may be something radically different from anything conceived of so far.

CHAPTER 11

The Imprint of Dark Energy on the Cosmic Web

While theorists are happily tromping through hypothesized extra dimensions, astronomers are scouring the familiar four dimensions of the observable Universe for the imprint of dark energy. Unlike dark matter, we have no ideas for how to create dark energy in an accelerator or how to build a detector to capture a small bit of this strange substance. And we can't even try to find a lump of it using gravitational lensing—dark energy doesn't clump like mass, but appears to more or less evenly permeate the cosmos. At present, our only hope of learning more about the major component of the Universe is to search for its impact on the architecture of the cosmos and the evolution of spacetime.

The structure of the Universe holds the first clue to the nature of dark energy. Most of the matter in the Universe is found in a network of sheets and filaments that extend through the Universe—a giant cosmic web of dark matter, dotted lightly with galaxies and punctuated by clusters, which perch at the intersection points of the web like giant cosmic spiders. This web is dark, but weighty—and its formation has been influenced by dark energy. The gravitational self-attraction of matter induces the collapse of dark matter and normal matter to form the galaxies, clusters, and strands of dark matter that wind and weave through the cosmos. Dark energy dilutes the ability of gravity to pull mass together and retards the growth of these structures. If we can obtain detailed measurements of the evolution of structure, we can extract the imprint of dark energy.

The acceleration of the Universe gives us a second window into the nature of dark energy. This acceleration increases the distances to key mile markers such as supernovae, quasars, and the cosmic microwave background (compared to a model of a flat Universe with no dark energy). We can look for ways to measure these distances, such as the supernova experiments discussed previously, and thus infer the detailed expansion history of the Universe.

In either case, the basic idea is to observe the Universe and then compare it to a suite of model universes created by a large computer simulation. Each model is based on a different recipe for the matter and energy content. These model universes will all look a little bit different—and the one that looks most like our observed Universe is the winner. A cosmological constant whose energy density is constant in time will produce a universe that can be distinguished from one that contains a time-varying quintessence field—the exact point at which the Universe switches from deceleration to acceleration will be different, for example.

We can also determine whether our model of gravity needs to be adjusted. Modified gravity models are likely to affect the expansion history and the formation of structure in a different way than dark energy does. The two cosmological probes of dark energy—changes in the expansion history of the Universe, and changes in when and how quickly structures were formed—are related to one another in a Universe filled with dark energy. If we measure the effect of dark energy on the expansion history, we can predict its impact on the growth of structure, and vice versa.

In modified gravity models, on the other hand, this relationship does not necessarily hold. The changes to the laws of gravity in these models affect the growth of structure in a way that cannot be inferred from the expansion history. We can no longer trace the expansion history and use this to predict the formation of structures. Thus, it is crucial to measure both the expansion history and the formation of structure—if these two observations don't agree on the same model of dark energy, we'll need to rethink gravity.

These effects are subtle, and collecting the data that can reveal the necessary details is challenging. But the stakes are so high that astronomers are enthusiastically expanding the hunt for dark energy in several directions, finding new ways to narrow down the list of suspects. The data we have in hand provide convincing evidence that dark energy exists, but it's a circumstantial case that has left us desperately looking for more clues. We have the equivalent of eyewitness accounts from three independent onlookers, each of whom caught a quick glimpse of an 8-foot 10-inch, 470-pound individual. At this point we're not even sure if the perpetrator is a member of any known species—and it seems likely that it is not.

UNIVERSE IN A BOX

In order to analyze the influence of dark energy, scientists are engaging the computational power of the world's largest supercomputers. They essentially create a universe in a box, a three-dimensional simulation of the Universe (which will be denoted by lowercase *universe* to distinguish it from our Universe), and watch as it evolves from shortly after the Big Bang to the present. Different cosmologies (different amounts or kinds of dark matter and dark energy, for example) can be dialed into the simulation, to create universes that can then be compared to the observed Universe. Given the characteristics of a real telescope or astronomical survey and its limitations (how bright does a quasar have to be to be seen through this telescope? how much of the sky will be included in the survey?), scientists can use the simulation to predict what the observers should see for a particular model. They count the numbers of clusters at different points in (virtual) cosmic history, hunt for the earliest galaxies, and trace out the dark matter web—and then rule out those models whose predictions are at odds with the data.

The largest cosmic simulation to date has been created by the

Virgo Consortium,[1] a team of astrophysicists from Germany, the United States, Canada, and the United Kingdom led by Volker Springel and Simon White at the Max Planck Institute for Astrophysics in Garching, Germany; and Carlos Frenk at Durham University in England. The Millennium Run simulation studies the dark matter in a cube of space that is 2.2 billion light-years by 2.2 billion light-years by 2.2 billion light-years, and follows its evolution over a 13.6-billion-year period of time that begins when the Universe was less than 12 million years old[2] and ends today. Normal matter in the form of gas is also added into the mix in order to follow the formation of galaxies and quasars, although in less detail than the dark matter.

The main ingredient in the recipe for making such a virtual model of the Universe is dark matter. The total amount of dark matter must be effectively digitized—broken down into individual miniclumps of dark matter that the computer is able to digest and process. Each miniclump represents many dark matter particles, since trying to follow all of the individual dark matter particles is far beyond the capability of current computers. Simulations with lots of miniclumps produce more detailed results than those with fewer, but the trade-off is in computing time: the more bits of matter the computer has to track, the longer it takes to complete the simulation. The Millennium Run simulation followed 10,077,696,000 bits of dark matter (the computer simulators refer to these as "particles," but they don't mean individual particles of dark matter), each with a mass of 1.2 billion solar masses. Thus, each "particle" of dark matter represents a huge number, of order 10^{67}, of dark matter particles (assuming that dark matter particles are typical cold dark matter particles known as WIMPs).

At the beginning of the simulation these miniclumps are distributed throughout the volume of the simulation in a manner consistent with the data from the cosmic microwave background—spread out in a fairly uniform, homogeneous manner, with only small deviations from a perfectly smooth density distribution. Then gravity is turned on. Gravity is the only force between the dark matter particles, and they move in response to their gravitational attraction to other particles. The universe is also set in motion, expanding according to the dictates of the particular cosmological model being studied. The computer looks at each bit of matter, analyzes the forces on it, and determines where it will move; then takes a step forward in time, rearranging the entire ensemble of matter bits with each step. The Millennium Run worked its way forward through over 13.6 billion years in 11,000 time steps, pausing to store the configuration of its universe 64 times during the simulation. The simulation generated almost 20 terabytes (a terabyte is 1,000 gigabytes, the unit usually used for laptop computer hard drives) of raw data, using 512 processors of an IBM parallel computer and 350,000 hours of processor time. In real time the program took just under a month to create one model of a universe with dark matter, dark energy, and a sprinkling of normal matter.

The results are stunningly beautiful (see Color Illustration 9). The upper half of panel (a) shows the distribution of normal matter; panel (b) is the corresponding web of dark matter. The brighter spots in both images indicate an increased mass density (they don't correspond to light intensity, which is zero everywhere in the dark matter panel). The very bright spot at the center of the images represents a rich cluster of galaxies. The galaxies and clusters of normal matter, panel (a), are found where the dark matter is most concentrated, and no galaxies are seen where there is no dark matter—but there are places with dark matter and no galaxies.

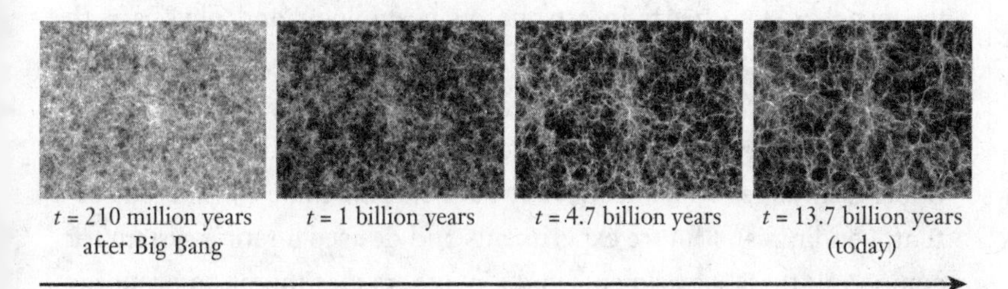

ILLUSTRATION 11.1 Time evolution of the cosmic web. Each panel represents a slice of a simulated universe, roughly 68 million light-years thick and 1.7 billion light-years across, viewed at four different times in the history of the universe. As gravity pulls mass together, the filaments and knots become more sharply defined and the cosmic web of matter comes into focus.

It's also possible to follow the evolution of a patch of the simulated universe over time. Illustration 11.1 follows the dark matter as it collapses from an amorphous, if slightly lumpy, distribution in a universe at the tender age of 210 million years, to the filamentary web structure seen in the universe of today, over 13.5 billion years later.

The details of this evolution reflect the impact of dark energy on the growth of structure. With access to a supercomputer, we can describe the dark matter web at any point in time for a given model of the Universe, in as much detail as our computing power will support. The real Universe is less accommodating. Astronomers need a way to map out the distribution of dark matter over the whole Universe, not just that which is bound in visible galaxies. And we need to be able to document how this distribution has changed over time if we hope to disentangle the effects of dark energy.

DARK ENERGY TASK FORCE

When a difficult problem of extraordinary urgency arises, a task force is often appointed to craft a solution and outline a way forward. Dark energy more than qualifies for such treatment. Three major U.S. funding agencies—the Department of Energy, the National Aeronautics and Space Administration, and the National Science Foundation—established the Dark Energy Task Force in 2005 and charged the panel of 13 experts in astronomy, cosmology, and physics with developing a strategy to optimize the search for the nature of dark energy.[3] The task force gathered ideas from the science community, laid out specific goals, and analyzed the main categories of observational techniques available to study dark energy. They considered current, pending, and future experiments and devised a rating system for comparing these experiments and any others that might be proposed in the future. In their final recommendations, they laid out a multistage approach to address the following key questions:

- Is a cosmological-constant solution to dark energy consistent with the data?

- If not, and the dark energy is not constant in time, how does it evolve?
- Is general relativity correct?

Experimentalists currently have four general methods for tracking down the effects of dark energy on the Universe. Each type of observation is designed to probe either the expansion history of the Universe or the growth of structure—or both.

Searching for Distant Supernovae

First on the list (historically) is the search for distant supernovae that was discussed in Chapter 10. This technique, which uses supernovae as standard candles (distant sources of light with known intrinsic brightness), measures the distance to a supernova as a function of its redshift (how fast the supernova and its home galaxy are receding from us) and thus provides information on the expansion history of the cosmos.

Mapping the Distribution of Galaxies

Next up are ripples in the distribution of galaxies across the sky. These ripples (known as baryon acoustic oscillations) are due to the same phenomenon that created the pattern of temperature variations in the cosmic microwave background. Sound waves in the primordial soup left their imprint on the cosmic microwave background in the form of the hot and cold spots that we see today, and these same waves also left their mark on the distribution of matter. In the latter case, these small ripples (overdensities) in the early distribution of normal matter translate into features in the pattern of how galaxies are spread throughout the Universe today.

If we make a detailed map of where all the galaxies are located on the sky, we can just barely make out spherical ripples where the density of galaxies is slightly higher than average. These rings of overdensity, not visible to the eye but detected by statistical analysis of the data, are the faint echoes of the primordial sound waves. They act as another kind of standard ruler—something whose intrinsic size is known—that we can use to determine distances. This standard ruler is actually more like a standard bubble—it's a three-dimensional spherical bubble whose diameter is known.

Each ripple is traced out by galaxies at a particular distance, and we can find this distance by comparing the observed size of the ripple or ring with the true size. Since the ripples in the galaxy density are due to the same sound waves that generate hot/cold spots in the microwave background, and we've measured the size of these spots, we can calculate that the true size (radius) of these rings (in the Universe today) is 466 million light-years.[4] The observed size will depend on how far away the ring is—the farther away an object is, the smaller it appears to be.

We also measure the recession velocities of the galaxies in the rings via their redshifts. The measurements of the distance to these standard rulers as a function of their redshift can be used to trace the expansion history of the Universe.

Surveying Clusters

Clusters of galaxies once again prove their cosmological worth via the third method for getting at the nature of dark energy. Large surveys of thousands or tens of thousands of clusters are sensitive to both the expansion history and the growth history of the Universe. Dark energy increases the distance to a particular redshift, and thus it increases the volume of the Universe between us and objects at that redshift. The number of clusters seen out to a particular redshift depends on the volume of space out to that redshift (for a given density of clusters, a larger volume will contain more clusters), and thus provides another measure of the expansion history.

The growth of structure is also very different in a Universe filled with dark energy. Once dark energy began to dominate the cosmic landscape, about 5 billion years ago, it suppressed the further growth of structures such as clusters. This means that in a dark energy Universe, the structure we see around us today must have already been in place much earlier—the effects of dark energy mean they can't have grown much recently.

This change in the growth of structure can be elicited from counting the number of clusters of a given mass, and determining how this number has evolved with time. Astronomers collect data on a huge number of clusters, then assign them to groups according to their redshift, which is an indicator of how far away and thus how far back in time they are. The clusters are then further subdivided according to how massive they are: the heaviest clusters go into one bin, those with slightly smaller masses into the next bin, and so on. The number of clusters of a given mass as a function of time traces the history of the growth of structures. A universe with dark energy will have more heavy clusters at an earlier time (a redshift of 1, for example) than will a universe without dark energy; and the exact numbers will depend on the specific model of dark energy.

Gravitational Lensing

The fourth technique is gravitational lensing. In the search for dark energy, the lens we are interested in is not a galaxy or a cluster, or even the halo of dark matter surrounding galaxies and clusters, but the entire web of dark matter that pervades the Universe. This invisible web is the largest gravitational lens in the Universe and will affect the path of light from every distant light source. Weak lensing of distant galaxies by the dark matter web will allow us to trace out the details of the bulk of the mass in the Universe. The observed pattern of weak lensing depends on the mass structure of this lens and its evolution (allowing lensing to probe the growth of structure), and the distances to the lens and the light source (which in turn depend on the expansion history). Lensing thus probes both the expansion history and the growth of structure.

All four of these techniques have known strengths and weaknesses. Supernova observations are well under way, and there is no question that detecting additional distant supernovae will improve our knowledge of dark energy parameters. This method will ultimately be limited by how well we understand these exploding stars and how precisely we can calibrate their brightness. Baryon acoustic oscillation techniques are still in the development stage, although in 2005 the

Sloan Digital Sky Survey reported the first observation of an extra bump in the distribution of galaxies in the nearby Universe due to baryon acoustic oscillations. The baryon acoustic oscillation technique has the advantage of using a very clean standard ruler (we understand the physics behind it), but it is less sensitive to dark energy than the other methods are. Surveys of clusters of galaxies add a probe of the growth of structure to the mix of dark energy search tools, but they require an understanding of how the hot gas in clusters behaves, which limits the accuracy of this method. Gravitational lensing requires accurate measurements of the distance to millions or billions of light sources, and the ability to determine the shapes of faint, distant galaxies to high precision. Nonetheless, gravitational lensing was highlighted by the Dark Energy Task Force as the method with "the greatest potential for constraining dark energy."

WEAK GRAVITATIONAL LENSING SURVEYS AND THE QUEST FOR DARK ENERGY

The ultimate lens of the Universe is the dark matter web. It stretches across spacetime in every direction, its tendrils becoming ever more diffuse as we look farther out into the Universe, further back in time, until eventually they dissolve into the lumpy soup of matter that fills the early cosmos. Light from distant galaxies (or any other distant source) must past through this web on its way to us, deflected by the tentacles and knots of mass in the web. Each photon of light follows a path defined by the way this lens warps spacetime, experiencing a series of tiny deflections as it dips into, out of, and around the undulations impressed on the cosmos by the dark matter web. The effects on the backdrop of galaxies are very subtle, but encoded within the pattern that we see through the lensing web is the outline of the web itself.

The goal of dark energy lensing experiments is to reconstruct this web so that we can study it carefully and extract the imprint of dark energy. We want to make a plaster cast of the footprints left behind by our suspect.

The basic process is the same as weak lensing induced by galaxies

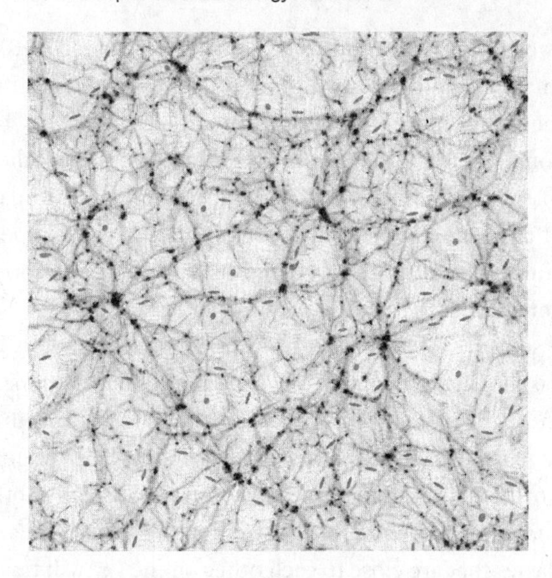

ILLUSTRATION 11.2 Lensing by the cosmic web. Scientists use computer simulations to study the lensing effects of the cosmic web by first creating a (simplified) background field of galaxies, then viewing these galaxies through a virtual realization of the cosmic web. The observed pattern—some galaxies are stretched into tiny ellipses while others remain nearly unchanged—allows scientists to reconstruct the architecture of the web.

and clusters, with two major differences. In all of the previous examples of lensing in this book, we've been able to approximate the lens as essentially two-dimensional. The "thickness" of a galaxy or a cluster (as we look through it) is so small compared to the distance to the lens or the light source that we can ignore it. A galaxy, for example, is roughly spherical (three-dimensional), but for the purposes of lensing we usually assume that all of the mass is projected onto a two-dimensional disk. Lensing measures this projected mass density. The assumption is that light from the source travels a huge distance more or less unimpeded (ignoring the effects of the cosmic web for the moment) until it encounters the lens; is bent by the warp in spacetime created by the lens; then travels unimpeded over another huge distance before we detect it in a telescope. The lens is located at a particular point in spacetime—its thickness along the direction the light is traveling can be ignored.

This is manifestly not so for the cosmic web, which extends all the way between any distant source of light and our telescope. We cannot approximate it as a single two-dimensional lens at a specific point in spacetime—and more important, we don't want to. The three-dimensional nature of the web is what allows us to track the evolution of the mass distribution and document the influence of dark energy. We want to get as complete a picture of this web as possible—the better our information on the web and its evolution, the more we'll be able to infer about the nature of dark energy.

The second major new feature is that the lens itself is invisible—the web is mostly dark matter. We can't point our telescope at a single direction in the sky and look for distortions in background galaxies around a particular cluster or galaxy, but must scan a huge section of the sky (or better yet the entire sky) and try to discern subtle patterns in the shapes and orientations of millions of distant galaxies. Light from galaxies that are close to each other on the sky will travel past the same parts of the dark matter web on their way to us, and thus the images of these sources should be distorted in similar ways, producing the kinds of patterns shown in Illustration 11.3. The precision with which we can discern these patterns—which in turn depends on how well we can measure the shape and distance of the background galaxies—will limit how much information about dark energy we can hope to extract from the lensing data.

Exactly how do we go about uncovering the dark matter web? Astronomers first make detailed observations of millions (the more the better) of faint galaxies. These galaxies are assumed to come in a variety of sizes and shapes, and they point in random directions. The data are then combed for correlations between the observed orientations of the galaxies—if the galaxies in a small patch of the sky show a slight preference to line up in the same direction or trace out a circular pattern, this pattern is most likely due to lensing.

Lensing of distant galaxies by the cosmic web was first announced by four independent groups of astronomers in 2000,[5] and it has been detected by many groups since then. With Einstein's Telescope, the effects of the strands of matter that stretch throughout the Universe have been seen.

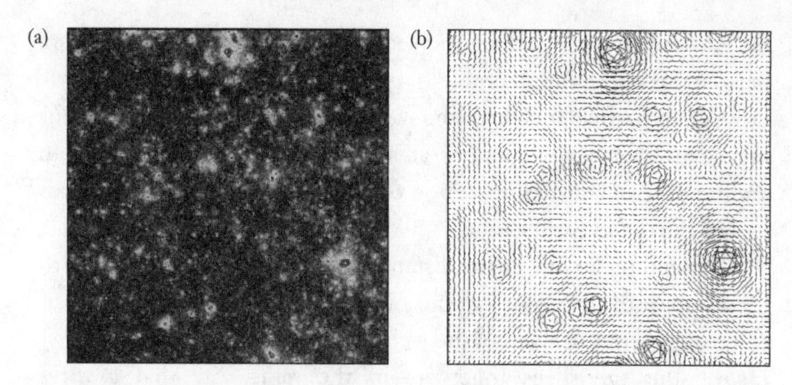

ILLUSTRATION 11.3 Cosmic shear. (a) The (projected) mass along the line of sight to distant galaxies in a simulated universe. (b) The effects of weak lensing by this mass distribution. The length of each line in this panel represents the strength of the lensing effect (how much a background galaxy would be stretched and magnified); the direction indicates the reorientation axis (the direction in which galaxies would tend to be stretched by the lensing). Even though the mass is not shown in panel (b), the observed shear pattern makes it easy to see where the large clumps of matter are located.

The pattern revealed by these detections is like an X-ray, which provides a flat, two-dimensional image—we see only the net result of the deflection of the light by all of the mass it has encountered on its way through the web. There is a lot of information in the images (more accurately, in the data—statistical methods are essential for extracting the patterns), but to study the time evolution of the web we need to create a full three-dimensional version. Instead of a two-dimensional X-ray, we want the equivalent of a CAT, or CT (computed tomography), scan of the Universe.

The light from the galaxies is lensed by all of the matter along its path, but not in a democratic fashion. In the case of a single massive object, gravitational lensing is most efficient when the lens is halfway between the source and the observer. For an extended web of mass, the part of the web that lies roughly halfway between the source and the observer makes the largest contribution to the lensing of the source. Each bit of mass that the light encounters will deflect it, but the light that eventually reaches us has been influenced more strongly by the mass at the halfway point. If we look through the web at background galaxies that are all at the same distance, the lensing pattern is produced for the most part by the section of the web that lies midway between us and this subsample of galaxies. We have an effective window into a particular slice of the web. The pattern produced by looking at another set of galaxies at a greater distance will reflect the architecture of a different section of the web—one farther away from us and thus at an earlier time.

Depending on how many distant galaxies we can see, we can divide them up into different groups according to their distance, and re-create a series of slices of the web in what is known as *lensing tomography*.[6] Put the slices together—in the same way that a three-dimensional model of the internal structure of the human body has been created in the Visible Human Project[7]—and you have a three-dimensional model of the dark matter web. The slices themselves can

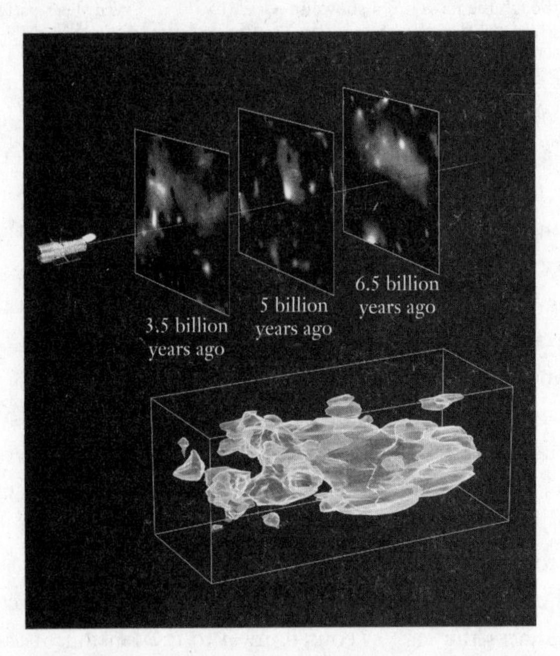

ILLUSTRATION 11.4 Dark matter tomography. The first three-dimensional scan of the dark matter web. The Hubble Space Telescope, as well as several ground-based telescopes, observed the gravitational lensing induced by different slices of the Universe in order to re-create the distribution of (mostly dark) matter. The section of the Universe mapped out covers an area roughly nine times that of the Moon, and extends billions of light-years out into space.

be compared (via the use of sophisticated statistical techniques) to the corresponding slices taken from computer simulations (see Illustration 11.1) to determine what universe—what model of dark energy—best matches the observed Universe.

Lensing tomography has already been road tested, and the first results were presented at the meeting of the American Astronomical Society in January 2007.[8] Shown in Illustration 11.4 is the first three-dimensional dark matter model of a section of the Universe. Using a combination of ground-based telescopes and the Hubble Space Telescope, the Cosmic Evolution Survey measured the shapes and distances of half a million galaxies to map the dark matter in a volume of space that covers almost 2 square degrees on the sky and stretches 6.5 billion light-years into the past.

CHALLENGING GENERAL RELATIVITY

The sensitivity of gravitational lensing to the evolution of structure makes it an essential part of any program to test general relativity. In a Universe where dark energy is the solution to the observed acceleration of space, the expansion history of the Universe and the growth-of-structure history can be directly related to one another. A precise measurement of the dark energy parameters from supernova experiments predicts what should be seen in gravitational lensing studies of the growth of structure and vice versa. Thus, measuring both of them provides a kind of cross-check. If the supernova data imply that the dark energy parameters are consistent with a cosmological constant, we expect that the evolution of the dark matter web as determined from gravitational lensing will also be consistent with a cosmological constant. Or if the lensing studies imply a particular quintessence solution, so should the supernovae. If the two kinds of probes don't agree, something is wrong with general relativity.

Modified gravity models do not have the same straightforward, one-to-one correspondence between the two measures. The expansion history reflects the behavior of (modified) gravity on the very large scale of the entire Universe; the growth of structure tests its actions on the much smaller scales of galaxies and clusters. In modified gravity models, gravity behaves differently on different scales—in essence there are more knobs to twiddle on a modified gravity model, so the two different probes (expansion versus growth) will not necessarily line up with one another.

We don't yet know if modifying gravity is a viable explanation for the observed acceleration of the Universe, but the experiments that are being designed to search for dark energy will put these models to the test.

CURRENT AND FUTURE EXPERIMENTS

The final recommendations of the Dark Energy Task Force laid out a plan for a series of experimental stages, moving from projects already in progress to future plans for large ground-based surveys and ambitious space-based missions. The panel devised a rating method to determine the merit of an experiment on the basis of its potential for ferreting out the parameters of dark energy: the equation of state that describes dark energy today (the value of w, the ratio of pressure to density described in Chapter 10) and how w has changed over time (if at all). If dark energy is a cosmological constant, w will be constant in time; if w is instead time-dependent, measuring this dependence can aid in narrowing down the list of quintessence models. The task force set a target for the scientific community, aiming for a tenfold increase in how well we know the product of these two key bits of information that characterize dark energy.

After careful analysis of the four main techniques and a host of current and future experiments that employ these techniques, the task force concluded that no single method is sufficient; the best approach is a combination of at least two of the four methods, one of which is able to probe the growth of structure. Furthermore, gravitational lensing was deemed an essential player.

Currently, or in the near future, the three cornerstone kinds of experiments that have documented the presence of dark energy—supernova searches, cosmic microwave background observations, and cluster mass measurements—are being continued or extended. There are supernova projects[9] that will each detect a few hundred nearby supernovae—the better we understand these local exploding stars, the more exact will be our measurements of the expansion of the Universe from their more distant cousins. The Planck satellite,[10] a mission of the European Space Agency scheduled for launch in 2008, will map the cosmic microwave background to an even higher precision than the current state-of-the-art Wilkinson Microwave Anisotropy Probe has done, allowing us to read the fine print on the earliest snapshot of the Universe. And more cluster surveys are in the works—surveys of broad areas of the sky that reach deeper into the cosmos to count, map, and weigh tens of thousands of clusters. The South Pole Telescope[11] located at the Amundsen–Scott research station is the largest telescope (10 meters across) ever to be deployed in the frigid, but astronomically advantageous, environs of the South Pole. It will scour the cosmic microwave background, looking for clusters using the Sunyaev–Zel'dovich effect—the tiny boost in temperature that the cosmic microwave background photons receive when they collide with the hot gas in clusters.

Current and near-future dark energy enterprises also include weak lensing and baryon acoustic oscillation surveys. The first detection of baryon acoustic oscillations by the Sloan Digital Sky Survey (SDSS) was based on less than half of the full data set that the SDSS will eventually have in hand, and a full analysis of all of the data should be available soon. Weak-lensing surveys are just beginning but have already yielded constraints on dark energy. For example, the Canada–France–Hawaii Telescope Legacy Survey will scan 170 square degrees (for comparison, the Moon covers an area of 0.2 square degrees); preliminary results from a subset of their data limit the dark energy parameter to $w < -0.8$.

CORNERING DARK ENERGY

The next generation of experiments, aiming to produce at least a threefold improvement in our description of dark energy, should begin to come online in the next decade.

Some are focusing on just one technique, such as baryon acoustic oscillations. These experiments will map the locations of galaxies—millions of galaxies. The SDSS on steroids, so to speak. Mapping the galaxies in two dimensions is not so hard—adding in the third dimension (obtaining distances from redshifts) is painful. The light from each galaxy must be collected individually in order to spread it out into a spectrum and determine the redshift. The SDSS is able to take spectra of 600 objects at a single time using optical fibers to isolate the light of each galaxy. The Wide-Field Multi-Object Spectrograph (WFMOS) experiment[12] proposed to up the ante by taking at least 4,000 spectra simultaneously.

Other proposals[13] incorporate three or four methods for constraining dark energy. The Dark Energy Survey is an international collaboration that has laid out a plan to survey 5,000 square degrees of the sky using the Blanco telescope, perched high in the Chilean Andes. The experiment will take images of the sky in four different colors and use this color information to estimate the redshift of each object—this method is not as precise as obtaining individual spectra, but it is a reasonably good approximation and much faster. Within their data they will hunt for the telltale pattern of baryon acoustic oscillations in the distribution of galaxies, collect cluster statistics, detect new supernovae, and measure the lensing effects of the cosmic web.

Other multitaskers include Pan-STARRS 4, an array of four 1.8-meter telescopes to be built in Hawaii; the ALPACA project, an 8-meter liquid mirror telescope that will be situated in Chile; and the One Degree Imager at the Kitt Peak National Observatory near Tucson, Arizona. Some of these experiments are fully funded, others are partway there, and some are still in negotiations with funding agencies and partnering institutions.

The really big guns are a bit further in the future. The exact details of the experiments that will be designed to achieve at least a tenfold increase in our ability to describe dark energy are likely to change before they see first light, but the general characteristics of what it will take to reach this goal are already mapped out. The stakes are high, and the projects are correspondingly bold and impressive—and expensive. The dark energy dream team is an ambitious trio of projects: the Large Synoptic Survey Telescope, the Joint Dark Energy Mission, and the Square Kilometer Array. Data from these dark energy programs, assuming they are funded and built according to current proposed schedules—an optimistic assumption—could be expected sometime in the middle of the next decade (2015 or later). All three include weak gravitational lensing as a key component.

Large Synoptic Survey Telescope

It may soon be possible to Google the Universe. The Large Synoptic Survey Telescope (LSST)[14] proposes to take astronomy to a whole new level, and Google is part of the team. *Synoptic* is defined as "constituting a general view of the whole of a subject." In this case, the subject is the Universe—or at least a chunk of it that covers half of the sky and extends from the Earth to galaxies over 11 billion light-years away. Under the direction of J. Anthony Tyson, an astronomer at the University of California at Davis, LSST will scan an entire hemisphere of the sky (about 20,000 square degrees) several times each month, generating on the order of 30,000 gigabytes of data every night. A 3-billion-pixel camera will be attached to a new 8.4-meter telescope to be built on a mountain in northern Chile (Cerro Pachon) and will collect images in six colors.

To aid in managing the enormous amount of data, Google has recently joined the 19 universities and national labs that have been working on the development of the LSST. The company behind the search engine that allows Web browsers to sift through the vast amount of information on the Internet will bring its expertise to the processing, storage, and manipulation of the unprecedented warehouse of astronomical information that LSST will generate. This vast database will itself be a target of astronomical research. Instead of collecting new data through a telescope, scientists will mine the overabundance of data stored by the LSST. Google will also be instrumental in developing a public access and search portal, allowing students and amateur astronomers an entry into this virtual universe.

The collaboration hopes to start operating the telescope in 2014 and run for 10 years. At the end of the program, it will have detailed observations of billions of galaxies and millions of supernovae. The stretching of spacetime will be exquisitely traced out by the supernovae, baryon acoustic oscillations will be clearly marked in the distribution of the galaxies—and the gravitational lensing power of the dark matter web will reveal its architecture in far more detail than ever before. Dark energy will have no place to hide.

If dark energy doesn't seem urgent enough, there is an added bonus. LSST can also act as a cosmic early warning system. It includes a repetitive imaging strategy that will allow it to identify nearby moving objects—near-Earth, a.k.a. "killer," asteroids—and track their orbits.

Joint Dark Energy Mission

The Joint Dark Energy Mission (JDEM)[15] takes the search for dark energy into space. The *joint* in JDEM refers to a joint undertaking of the U.S. Department of Energy (which funds significant research in fundamental particle physics, including support of national labs such as Fermilab) and NASA. Until late 2008 there were three independent proposals competing for the design of JDEM; however, the competition was abruptly closed, and elements of all the proposals (Destiny, SNAP, and ADEPT) are now being considered as part of a common reference design. The final design may include all three of the key dark energy search techniques—baryon acoustic oscillations, supernovae observations, and weak lensing. Destiny is a space telescope that will focus on both observing distant supernovae and gravitational lensing; the SuperNova/Acceleration Probe (SNAP) proposes to employ a 2-meter optical telescope to measure supernovae, baryon acoustic oscillations, and weak lensing; and the Advanced Dark Energy Physics Telescope (ADEPT) will concentrate primarily on baryon acoustic oscillations.

Space-based telescopes offer several advantages for weak-lensing surveys. When a point of light (such as a star) is viewed through a telescope, the result is never a point of light. Instead, the light is smeared out a bit—a point of light appears as a little smudge of light that is not even perfectly round. The main cause of this smearing in a ground-based telescope is the Earth's atmosphere; moving the telescope into space greatly reduces this blurring effect. Lensing studies are looking for tiny (about 1%) distortions of galaxy shapes due to lensing by the dark matter web, so any additional smearing by the telescope or the atmosphere makes teasing out this small effect much more difficult. In space, the smearing is minimized, increasing the number of faint galaxies whose shapes can be discerned with high enough precision to contribute to the lensing experiment, and the accuracy with which the galaxy shapes can be measured. Observations in space also increase the depth of a survey: galaxies are visible at greater distances, thereby increasing the lensing signal due to the cumulative lensing of a longer path through the dark matter web, and allowing us to probe more distant reaches of the web.

Square Kilometer Array

The third player in the future dark energy program is firmly planted back on Earth. The Square Kilometer Array (SKA)[16] is not a single telescope, but a large arrangement of thousands of radio telescopes with a combined collecting area of 1 square kilometer—almost 250 acres. The telescopes, which look like large versions of the television satellite disk that may be attached to your house, are actually radio antennas whose signals can be digitally combined to simulate one large telescope. Each individual antenna dish will be 30 to 40 feet in diameter, but together they create an effective telescope whose diameter is equal to the largest separation between two antennas—over 1,800 miles. (The collecting area of a telescope is a guide to how much light it collects; the effective diameter is related to the resolving power of the telescope—the amount of detail it can discern in distant objects.) The design details are still in the development phase but the antennas will be spread across thousands of miles, with a central core a few miles across containing about half of the receivers. SKA will most likely be located in either South Africa or Australia.

SKA will be searching for radio signals from a wide range of sources, including dust rings (the precursors of planets) around newly formed stars and pulsars (spinning neutron stars). But the element of SKA of interest to dark energy searches is its ability to detect hydrogen gas at great distances. A gas of neutral hydrogen (a hydrogen atom that has a firm grip on its electron) emits a particular frequency of light that can be detected by radio telescopes. This light signal, which is emitted at a wavelength of 21 centimeters (a frequency of 1.4 gigahertz), acts as a homing beacon for radio telescopes. The larger the diameter of the radio telescope, the more precisely the signal can be located on the sky. Furthermore, the hydrogen light signal is redshifted by the expansion of the Universe, so the exact frequency at which the signal is detected by the telescope pinpoints its location in the third direction (away from us). SKA will provide an exquisitely detailed, three-dimensional map of the neutral hydrogen in the Universe.

This emission line is extremely difficult to detect because the Earth's atmosphere and the transmission of television programs can run interference with our ability to pick up the faint signal of distant hydrogen, but the prize at the bottom of the box is well worth digging for. Galaxies are full of neutral hydrogen gas, and even when the dust in a distant galaxy blocks the visible light of its stars, the 1.4-gigahertz hydrogen signal gets through. Furthermore, using this technique we can see galaxies even before they light up with stars—the hydrogen gas in an early galaxy is there before the stars turn on. This signal will allow SKA to detect on the order of a billion galaxies that can be used in measuring baryon acoustic oscillations over a large stretch of cosmic history.

In addition to the emission line, galaxies emit a more general glow of radio light, which can be detected by SKA and used as sources of light for gravitational lensing. Detailed maps of these faint radio sources—their positions, shapes, and distances (redshifts)—will bear the imprint of lensing by the dark matter web.[17]

DARK ENERGY UNVEILED

If the megaprojects described in the previous section are funded and built, we should have the answers to the questions outlined by the task force within the next two decades. LSST could be gathering data by 2014; JDEM could be launched by the middle of the coming decade; and the first phase of SKA could be coming on line at about the same time, although this depends on funding. Each of the programs has its own strengths and weaknesses, and there are still many details (precisely how well we can measure the shapes of very faint galaxies) that won't be known until the data start to come in. Scientists are working on a better understanding of the telescopes and improved methods for gathering the data, but each technique will always have limitations. This is why the task force is emphasizing multiple assaults on dark energy at every stage.

Our elusive subject will now be tracked down by three new detectives, each with a unique tool kit for uncovering evidence, discovering new clues, and refining the picture of dark energy. Their combined resources will provide a much clearer description of the major component of the Universe, and the parameters of dark energy will be known to an accuracy of a few percent. We'll be able to trace how the dark energy changes over time and determine whether the observed acceleration of the Universe can be explained with a cosmological constant. And we'll know if the solution to the riddle of dark energy requires a major modification of Einstein's theory of general relativity—or if a strange new substance fills the cosmos.

CHAPTER 12

Gravity Waves

As more detailed and ambitious windows into the far reaches of space are opened, Einstein's Telescope is going to be increasingly important in our quest to understand the Universe. But it won't always be helpful. The lensing of distant light sources by the cosmic web allows us to study the distribution of dark matter and look for the imprint of dark energy, but it can also blur our view of the early Universe. And we desperately want to look back to ever-earlier epochs in cosmic history. The Big Bang theory has been tremendously successful so far, but it doesn't explain everything. Why does the Universe look so much the same in all directions? If galaxies and clusters grew under the influence of gravity from small perturbations in the otherwise smooth Universe, what produced these tiny ripples in the density of matter? And can we even imagine what came before the Big Bang?

The last question remains a topic of much debate and will not be answered anytime soon. But the further back in time we can push our understanding of the cosmos—the deeper our understanding of the fundamental nature of space and time, matter and energy—the more we might hope to explore ideas that take us beyond the beginning of our own Universe.

Gravitational lensing is instrumental for completing our census of matter and energy, but it may also undermine our ability to ferret out the dynamics of the nascent Universe in the first critical fraction of a second of its existence. We can't see into this era directly with any kind of telescope, nor can we re-create the high-energy conditions of this period in any of our particle accelerators. The best we can do is to look as far back in time as possible, and search the data we collect for hints of even earlier times.

Our earliest image of the Universe comes from the most distant source of light—the cosmic microwave background—which was generated when the Universe was 380,000 years old. To go beyond this point in time, we use our understanding of nuclear and particle physics and observations of the abundance of the light elements (such as helium and lithium) as a window into the first minutes of the cosmos. But even this is not far enough. We need to find a way to probe the era when the Universe was only about 10^{-32} second old.

The cosmic microwave background—the afterglow of the Big Bang—may be our best bet for gleaning information about the most distant past. Though the cosmic microwave background itself was created hundreds of thousands of years after the Big Bang, it holds critical clues to a much earlier epoch, if only we can get at them. A gold mine of information about the Universe and its composition, the microwave background light has already answered cosmological questions that have been debated for years, and we know we have not yet exhausted its riches. Embedded within the fine print of this ancient light are the gravitational echoes of the first moments in time.

Extracting signals from the earliest epoch of the Universe will require extremely precise observations of the microwave background. In this endeavor, Einstein's Telescope becomes a problem. The web of dark matter acts as a screen through which the microwave background is viewed, and the lensing induced by this irremovable screen distorts the images of the cosmic microwave background in ways that may mask the subtle clues we so anxiously seek. We need to understand the lensing properties of the cosmic web of matter in order to subtract its effects, so that we can peer into the first moments of cosmic history.

A SMOOTH UNIVERSE

The cosmic microwave background provides some of the strongest support for the Big Bang model of the Universe—and at the same time points out one of its weaknesses. The Big Bang theory describes the expansion and evolution of the Universe from an initial extremely hot and dense state to the cold and sparsely populated Universe of the present era, and one of the first predictions of this theory was the existence of the cosmic microwave background. Early proponents of the Big Bang model calculated that light from the hot, dense primordial soup will cool as the Universe expands and be seen today as a microwave radiation that fills the Universe at a temperature of a few degrees above absolute zero (early estimates ranged from about 5 to 50 degrees).[1] The detection of a 3-degree microwave background by Penzias and Wilson in 1965 lent strong support to the Big Bang model. Less than 30 years later the COBE satellite map of the entire microwave sky revealed that the temperature of this background is exceptionally uniform. Scan the sky with a microwave telescope and you'll find that the temperature of the Universe is 2.725 degrees above zero everywhere you look.

This is a problem. Such an even-temperatured Universe[2] is hard to understand within the context of the standard Big Bang theory. How do different parts of the Universe come to agree on a single temperature? Information, such as temperature, is not transmitted instantaneously from one part of the Universe to another. It takes time for information to travel—at best it can move at the speed of light, but no faster. This means that different parts of the Universe—those that are separated by a distance greater than the distance light could have traveled since the Big Bang (the beginning of time)—have had no way to exchange information with one another.

The general problem is familiar to anyone who has ever stayed in the bathtub too long. The water cools off, so a bit more hot water from the tap is added to warm it up again. As the hot water pours into one end of the tub, the distant end remains cool, at least initially. Given enough time, the temperature will even out as the water at the hot end comes into contact with cooler water next to it, transferring heat from hot to cold regions until an equilibrium temperature is reached. The speed of this process is limited by the rate at which heat diffuses throughout the water in the tub.

In the Universe, communication between different regions is limited by the speed of light. Some parts of the Universe that we can see today are separated by a distance greater than the distance that light could have traveled since the beginning of time. In the standard Big Bang theory, these widely separated regions, which we observe to be at the same temperature, have had no communication with one another—there is no way for them to have reached a common temperature (unless they started out that way).

Scientists refer to this as the *horizon problem*. At any point in space, there is a horizon defined by the distance that light has traveled over the age of the Universe. Light from anywhere within the volume of space demarked by this horizon has had time to reach the central point. As the Universe ages, the size of the horizon grows. If we neglect the expansion of the Universe for the moment, then for each year older, the horizon size is one light-year bigger.[3] This increase brings into contact regions of the cosmos that—in the standard picture of the Big Bang—have never before had an opportunity to exchange information.

By the same token, the existence of this horizon also implies that as we sit in the center of our horizon and look out at the Universe, we can see regions that were not in contact with one another when the light that we detect was emitted, as depicted in Illustration 12.1. In the standard Big Bang theory, the cosmic microwave background light arriving at our telescopes from one direction has never been in contact with the microwave light coming from the opposite direction on the sky. So how can the temperature of the light arriving from any and all directions be synchronized so perfectly?

The Big Bang theory simply cannot explain the smoothness of the observed Universe. One way out of this conundrum is to assume that the Universe was completely smooth and homogeneous to begin with. This kind of solution is not very satisfying to scientists. It simply

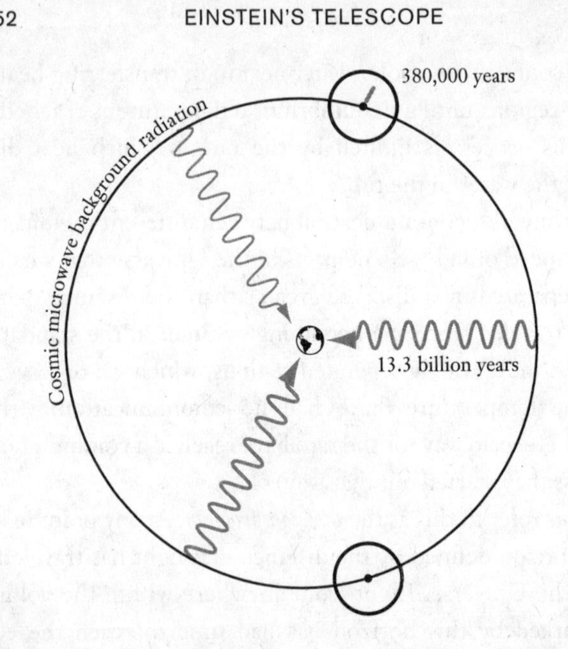

ILLUSTRATION 12.1 The cosmic horizon. The distance light can have traveled since the beginning of the Universe defines our current cosmic horizon. Ancient light from the Big Bang arrives at the Earth today from all directions, having traveled over 13 billion years from the time it last interacted with the particles in the primordial soup. This light has an extremely uniform temperature of 2.725 degrees above absolute zero, yet at the time the light was emitted the cosmic horizon was much smaller, since the Universe was less than 400,000 years old. This raises the question of how widely separated regions of the Universe (which had no way to communicate with one another at the time the light was emitted) could be at exactly the same temperature.

assumes that this is the way things are, instead of providing a physical explanation for why. It doesn't add to our understanding, or make any predictions about what else we should observe.

In 1979, a young physicist named Alan Guth came up with an ingenious addition to the Big Bang model that resolves the horizon problem, predicts that the Universe should be flat, and, as an added bonus, produces ripples in the distribution of matter that lead to the formation of structure.[4] He wasn't originally trying to achieve any of these results. A particle physicist intrigued by the possibility of unifying the forces of nature, Guth wanted to find a way around a generic prediction of so-called grand unified theories—the production of copious numbers of unwanted particles called *magnetic monopoles*.

Any magnet you might find in your house has two ends: a north pole and a south pole. Cut the magnet in half and you produce two new magnets, each with a north and south pole. No one has ever seen, or been able to manufacture, a magnet with just one pole—including a "magnet" composed of a single particle. Elementary particles that are the equivalent of a magnet with only one pole are known as magnetic monopoles, and no one has ever detected one of these either. Astronomers have searched the Universe for these strange particles and come up empty-handed.

Why worry about them then? Particle physics models that unify the electromagnetic, weak nuclear, and strong nuclear forces generically predict that a huge number of magnetic monopoles will be produced in the Big Bang. Such models are an essential step in the quest to unify all of the forces of nature, and we need to understand why no monopoles have been found if we want to continue to consider these kinds of models as part of a grand theory of everything.

INFLATION

Guth suggested a period of inflation—not of the dollar, but of space. The basic idea is that at a very early point in cosmic history, the Universe experienced a brief period of accelerated expansion, doubling its size every 10^{-37} second or so. This burst of expansion was over quickly—lasting only a minuscule fraction of a second—and by the time the Universe was 10^{-32} second old, inflation had ended and space had settled down into the more leisurely expansion of the standard Big Bang model.

Inflation doesn't refer to a specific model—a wide variety of models predict a period of exponential expansion in the early Universe, and the list continues to grow. The general mechanism behind many of these models is identical to the physics that powers the accelerating Universe in quintessence models (discussed in Chapter 10). The early Universe is bathed in an energy field—usually called the *inflaton field*—that is somehow held above its minimum energy value.

Our toy model of a marble perched on the edge of a well is applicable here (see Illustration 10.5 on page 217). As long as the marble is above the bottom of the well, it has potential energy. In models of inflation, the inflaton field is kept in a state (known as a *false vacuum*) where its potential energy is not zero. This potential energy corresponds to an energy of the vacuum, which then powers a period of exponential expansion of space. When the inflaton field reaches its minimum level, inflation will end. The original potential energy of the field is ultimately converted into the hot primordial soup, and all of the matter and radiation that fills the Universe is created in this transition. The standard Big Bang story starts from here.

During inflation, the Universe expands in size by a factor of 10^{50}—or more. (For comparison, without inflation it would have increased by less than a factor of 100 during the same time period.) Therefore, the

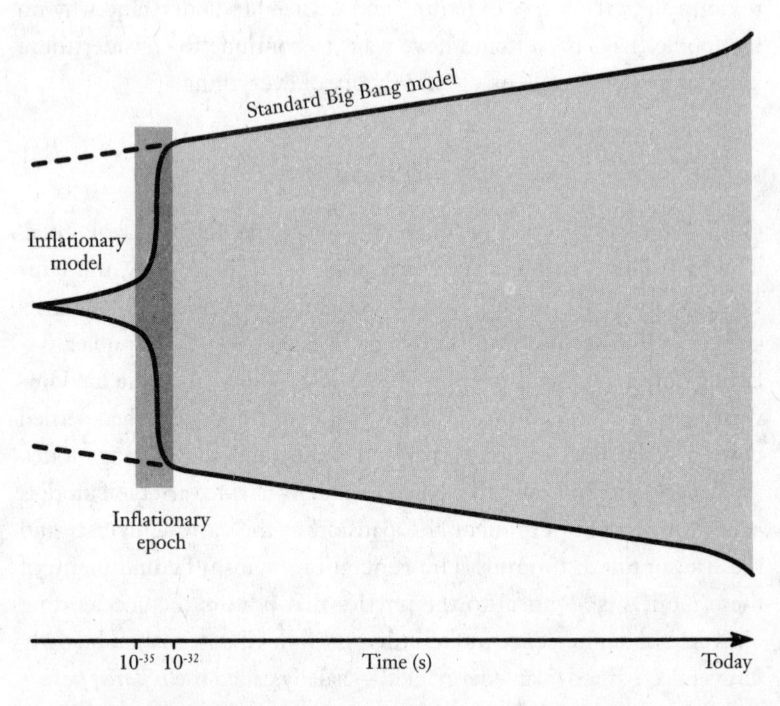

ILLUSTRATION 12.2 Inflation. According to the theory of inflation, the Universe went through an early period of extremely rapid growth, experiencing an exponential increase in volume over a very short period of time, before settling down to the more leisurely expansion of the standard Big Bang model.

part of the Universe we see today, and probably a much larger region, was all part of the same tiny patch of the Universe 13.7 billion years ago. Start with the size of the observable Universe today, which has a diameter of almost 10^{27} meters. Now reverse the Big Bang expansion of the past 13.7 billion years, then shrink by another 50 orders of magnitude or so, and the end result is a patch roughly 10^{-50} meter across. Everything we can see today was contained within this tiny patch.

The preinflation size of this patch is smaller than the horizon size—the distance light could have traveled—of that time period, and thus what is now the observable Universe was all within the same cosmic horizon, and therefore able to reach a common temperature and composition. During inflation, areas that are close neighbors within this initial region rush apart so fast and so far that by the end of the inflationary period they are far outside each other's horizons and no longer able to communicate with one another.

It will be billions of years before some of them reacquaint themselves. As the Universe expands after inflation, the horizon size also grows, bringing back into contact former neighbors. When we look around at the Universe today, we're seeing these reunited regions—and they all look the same because they came from the same initial patch. The horizon problem is solved.

Inflation also very nicely solves the magnetic monopole problem. After inflation, the number of any kind of particle (including monopoles) in the original patch is diluted away to essentially nothing. Assume you have 100 Ping-Pong balls in a box whose size is 3 feet on a side. Now increase the size of the box so that each side is 93 million miles across (roughly the distance between the Earth and the Sun). Good luck finding even one of the 100 balls. In any volume of space the same size as the original box, or 100 or even 100,000 times larger, there is likely to be at most one Ping-Pong ball, and probably none.

The Universe inflated much more than this, such that the entire observable region of the Universe today will contain at most one magnetic monopole. In fact, inflation dilutes the density of any kind of particle to less than one per observable Universe. The particles that exist today (or the particles from which these particles were produced) were all created at the end of the inflation from the energy of the inflaton field. Monopoles are too heavy to be produced in this process, and thus they have no way to repopulate the cosmos.

As Guth was formulating this solution to the monopole problem, he recognized that his theory of inflation could explain another long-standing puzzle in cosmology. Why is the Universe so flat? In the standard Big Bang model, a universe that is not precisely, perfectly flat at the time of the Big Bang will become more and more curved over time. Any minuscule amount of initial curvature will be magnified. This presents what is known as a fine-tuning problem. In order to appear so flat to us today, the Universe must have been either exactly flat at the time of the Big Bang, or at least much, much closer to exactly flat than it is now. Again, we either have to assume that this is just the way it started—or find some physical mechanism that can explain why.

Highlighted at the top of a page in Guth's 1979 notebook from this period is the phrase "SPECTACULAR REALIZATION."[5] His excitement was justified—he had just realized that his new theory of inflation predicted a flat Universe. Picture a caterpillar on the surface of a small balloon—the curvature of the balloon will be obvious to the bug (assuming it's able to process such information). Now inflate the balloon to the size of the Earth—looking around, the bug will surmise that the surface on which it's inching along is flat. Similarly, after a period of inflation, space expands exponentially so that any initial curvature of the Universe is effectively erased—the curvature is inflated away. The Universe we observe will appear to be flat, even if it began with an intrinsic curvature.

Inflation thus provides an elegant explanation for why we observe a flat, homogeneous, and monopole-free Universe. It also comes with a bonus feature: during the period of inflation, ripples in the density of the primordial soup are created. These ripples are critical to the formation of the Universe we see today.

The energy field that powers inflation is subject to the rules of quantum mechanics, so it can't have exactly the same value at exactly the same time everywhere in space. There will be tiny fluctuations in the value of the field on the very small scales where quantum mechanics becomes important. In our toy model of a marble rolling into a well, picture a marble/well setup at every point in space (an essentially infinite number of marbles and wells). Quantum mechanics states that the marbles will not all roll down at exactly the same speed—the uncertainty principle insists on random variations in this speed at the quantum level. Thus, the value of the inflaton field will be slightly different at different points in space, creating a pattern of subtle spatial variations.

At the end of inflation, the energy of the inflaton field is transformed into new particles, and the quantum fluctuations in the inflaton field are translated into density perturbations in the newly created particle soup. These, in turn, seed the formation of galaxies. The gorgeous views of the Universe seen through the Hubble Space Telescope are courtesy of quantum mechanics in the first moments of time.

So far, inflation has passed all of the observational tests to which it's been subjected. The measured temperature variations in the cosmic microwave background are consistent with the predictions of inflation, as is the observed geometry of the Universe inferred from this data. But we don't yet know for sure if inflation really happened. Furthermore, many models of inflation satisfy the observational constraints, each motivated by a different picture of the physics in the early Universe. Some involve multiple energy fields; others are embedded within the context of strings and branes. Our best hope of differentiating models, and further strengthening the inflationary paradigm, is the next generation of cosmic microwave background experiments.

In theory, the cosmic microwave background bears witness to the physics of the inflationary epoch. During inflation, the same quantum mechanical processes that lead to density perturbations in the distribution of matter and energy also create tiny disturbances in spacetime, creating waves of gravity that ripple across the cosmos. The transition period at the end of inflation may add to these oscillations. The process by which the energy of the inflaton field is transformed into a hot soup of particles may be far from gentle, as clumps of matter are produced and collide with one another, further stirring up spacetime. In a sense, the birth pangs of the Universe perturb spacetime, and these perturbations travel throughout the Universe as waves of gravity.

The predicted strength of the gravity waves produced by inflation depends on the specific model of inflation under consideration. In general, the earlier inflation occurs,[6] the higher the amplitude of the gravity waves will be. These gravity waves may be too faint to be detected by any method we have yet conceived of. Or they may leave an imprint on the cosmic microwave background—a faint echo of the epoch of inflation.

GRAVITY WAVES

Einstein's theory of general relativity is not a static theory, but a dynamic description of the Universe. Objects such as stars, planets, and galaxies move through spacetime—and spacetime shifts in response. And just as it's possible to generate waves in water, it's possible to create waves of spacetime. Gravity waves are curves and bumps in spacetime that propagate at the speed of light, moving through the Universe unaffected by anything they pass through. Spacetime itself is being jiggled by the motion of the matter within it.

The amplitude of gravitational waves is generally extremely small, and falls off as the wave travels farther away from the source. For humans this is probably a good thing—we don't need to worry that the Earth will be jostled by the gravity waves from a distant source (two black holes whirling around each other, for example), causing us to bob around like a small raft in the wake of a large speedboat. But we're confident that gravity waves are large enough that we can hope to detect at least some of them.

The cosmos is filled with gravity waves from a variety of sources in addition to inflation. Very massive objects moving quickly will swizzle the space around them, generating waves that propagate outward. Colliding black holes do a nice job of stirring up spacetime, and they may be our best bet for directly detecting gravity waves. Two neutron stars spiraling around one another are another good source. Such a system led to the first indirect proof of the existence of gravitational waves.

In 1993, Joseph Taylor and Russell Hulse were awarded the Nobel

ILLUSTRATION 12.3 Ripples in spacetime. An artist's conception of the gravitational waves created by the motion of two orbiting black holes. The waves correspond to ripples in the fabric of spacetime, which propagate outward from the black hole pair as they whirl around each other—headed for an eventual cataclysmic collision.

Prize in Physics for their 1974 discovery of two neutron stars locked in a tight orbit about one another.[7] One of the neutron stars is a *pulsar*—a rapidly spinning neutron star that emits a regular pulse of light, much like the beacon in a lighthouse. The timing of the pulses (which are emitted in the radio wavelengths of light) from this particular pulsar is very regular, creating in effect a high-precision clock. As the pulsar orbits its companion, the regular ticks of this pulsar clock are Doppler shifted. We detect a series of light pulses spaced closer together (as the pulsar swings toward us), then farther apart (as it moves away), then closer together again in a regular pattern as the pulsar moves in its orbit about the other neutron star. This allows us to very carefully time the orbital period of the system.

This binary system, known as PSR 1913+16, provides a perfect testing site for general relativity. Each neutron star has a mass of about one and a half Suns packed into a sphere roughly 13 miles across. They're separated by a few times the distance between the Earth and the Moon, and they orbit around each other every 8 hours. According to Einstein, the motion of these neutron stars should generate gravity waves, which will carry energy away from the system. And as energy from the system is radiated away in the form of gravity waves, the orbital energy will decrease and the neutron stars will spiral in toward one another, on an increasingly tighter and faster orbit. After 4 years of careful observation, measurements of the orbital decay (the decrease in orbital period) of PSR 1913+16 were published. The data were in excellent agreement with the predictions of general relativity, offering convincing evidence of the existence of gravity waves. Current measurements of this system agree with Einstein's theory to better than one-half of 1%.

Direct detection of gravitational waves is next on the list, and the search is already under way. Several highly sensitive gravity wave detectors, such as the Laser Interferometer Gravitational-Wave Observatory (LIGO),[8] have been constructed to monitor spacetime, looking for the characteristic signal of a passing gravity wave. LIGO is built like a large *L*, with two arms, each 2.5 miles (4 kilometers) long, arranged perpendicular to one another. Two masses are hung from each arm, one near the vertex of the *L* and the other at the far end, and the distance along the arm is measured by a laser beam bounced back and forth between the two masses. The laser light from one arm is then combined with the beam from the other arm to create an interference pattern—a pattern of bright and dark spots. If the distances along each arm remain exactly the same, the laser beams will remain in phase with one another and the interference pattern will be unchanged. If the distances change, the beams will be thrown out of phase and the resulting interference pattern will be altered.

As a gravity wave travels through the Earth and the LIGO detectors, it will stretch space in one direction while simultaneously compressing it in the perpendicular direction, changing the distance between the two masses on each arm by a tiny amount. The resulting changes in the interference pattern of the laser light will signal that a gravity wave has just passed by.

Using this technique, LIGO should be able to detect a change in the 2.5-mile length of each arm on the order of 10^{-17} inch—much less than the size of a hydrogen atom. This level of precision requires that the experimental setup be enclosed in a vacuum, so that the motion of air molecules doesn't affect the results. It also requires that there be more than one detector. Earthquakes, passing trucks, and other unwanted rumbles will generate a signal in the detector that can mimic or swamp the signal from a passing gravity wave, but all of these effects are relatively local to a specific place on Earth. A truck rumbling down the highway in California will not register in a seismic counter in Florida.

A passing gravity wave is not local—it will hit the entire Earth at essentially the same time. To distinguish between uninteresting noise and a gravity wave, LIGO has built two detectors spaced 2,000 miles apart—one in central Louisiana, the other in southern Washington. Any detection not picked up by both installations at the same time is not considered a candidate gravity wave. Two similar detectors have been constructed in Europe: GEO600, a British–German experiment located in Germany; and Virgo, an Italian–French detector constructed in Italy. All three detectors are coordinating their efforts, but so far no detections have been made.

One way to eliminate all of the unwanted noise from Earth-based sources is to put a gravity wave detector out in space. The Laser Interferometer Space Antenna (LISA)[9] is a joint effort by NASA and its European equivalent, the European Space Agency; it is scheduled for launch sometime after 2015. LISA consists of three spacecraft arranged in a giant triangle roughly 3 million miles on a side. Beams of laser light will be bounced back and forth between the detectors on each ship in order to measure distances between them to high accuracy, again using interferometry techniques. LISA will be sensitive to lower-frequency gravity waves than LIGO and other Earth-bound detectors are, providing a broad window into the realm of gravitational radiation. Scientists hope to directly detect the gravity waves created by pairs of black holes or neutron stars in the Milky Way, or the collision and merger of supermassive black holes in distant galaxies.

However, neither LIGO nor LISA has the sensitivity to detect the gravitational wave background—the faint oscillations of spacetime created in the first moments of the cosmic history—that is predicted in most models of inflation. A next-generation space experiment (the Big Bang Observer) has been proposed, but it's unlikely to be built any time soon. The best detector currently at our disposal for finding the gravity waves from the early Universe may be the cosmic microwave background.

THE SEARCH FOR ANCIENT WAVES OF GRAVITY

Gravity waves from the era of inflation carry with them details from times far earlier than we can probe with light, but we may need to use light to detect them. Gravity wave detectors constructed on Earth search for the characteristic stretching and squeezing of spacetime induced by a passing gravity wave by carefully monitoring the distance between two masses. In effect, they are looking for the motion of these masses caused by the gravity wave, just as we can watch a buoy bob up and down as a water wave passes by. Nature has provided a similar detector in the cosmic microwave background. As gravity waves from inflation pass through the primordial soup, they jiggle it, creating waves in the density of matter and radiation. These motions are imprinted on the light as it separates from the matter, and their presence is encoded in the fine details of the microwave background we observe today.

Extracting the information will not be easy. The amplitude of gravity waves from the early Universe may be too small to be detected in any foreseeable cosmic microwave background experiment. Even under the most optimistic inflationary scenarios, the effects of gravity waves on the microwave background are swamped by the contributions of the density perturbations that were discussed earlier. And the most promising signal of gravity waves in the cosmic microwave background may be obscured by the lensing power of the dark matter web.

Detecting the imprint of gravity waves in the cosmic microwave background requires mapping the pattern of hot and cold spots at higher precision than has yet been achieved. The Planck satellite, launched in 2009, is designed to provide this level of precision. In addition to measuring the variations in the temperature of the cosmic microwave background light, Planck will map the variations in the polarization of the light.

Polarized light is familiar to anyone who has donned Polaroid sunglasses. A light wave is an electromagnetic wave that oscillates in a plane perpendicular to the direction of the light beam. Unpolarized light, such as that from the Sun, oscillates in all directions on this plane. Polarized light has a preferred direction. Light that is reflected off of a surface, such as the pavement on a highway, is polarized (at least partially)—the light waves oscillate back and forth only in the direction parallel to the surface. Polaroid sunglasses cut down on the glare from the road by selectively blocking light waves that oscillate in the direction parallel to the road.

The light from the cosmic microwave background is also scattered just before it completely decouples from matter. This last scattering, as the light bounces off of electrons in the primordial soup, partially polarizes the light. The electrons are hit with light coming from all sides, and because of tiny temperature variations in the primordial soup, light coming from one direction can have a slightly different temperature (energy) than the light from another direction. If the temperature of the light that strikes an electron from one direction is different from the temperature of the light coming in from the perpendicular direction, the scattered light will be polarized.[10]

These temperature differences arise from two main sources: matter density fluctuations (which correspond to hot and cold spots as discussed earlier) and gravity waves. As gravity waves move through the sea of cosmic microwave background photons (light particles), they stretch and squeeze spacetime, and the wavelength of the photons is stretched and compressed in the process. Longer wavelengths correspond to lower energies and thus lower temperatures.

Only a small fraction of the cosmic microwave background light is polarized, so the strength of the polarization signal is even smaller than the observed temperature variations. Nonetheless, in 2002 the Degree Angular Scale Interferometer (DASI)[11] collaboration reported the first detection of polarization of the cosmic microwave background. The signal they found was due to the polarization induced by matter density fluctuations, and it was in perfect agreement with the prediction based on the observed temperature variations of the cosmic microwave background (which are created by the same matter density fluctuations).

The polarization signal expected from gravity waves is even fainter and harder to detect. Through careful analysis of a higher-resolution version of the polarization map (which the Planck satellite should be able to provide), the component due to only the gravity waves can be extracted—in theory. In practice, gravitational lensing of the cosmic microwave background will interfere. Gravitational lensing will create patterns on the cosmic microwave background similar to those imprinted by gravity waves, and disentangling the two signals will be challenging.

CORRECTIVE LENSES

When the Hubble Space Telescope was first put into orbit in 1990, astronomers soon realized that the primary mirror in the telescope was imperfect. It was ground a little too flat at the edges, distorting the images collected by the telescope. (The mirror has a spherical aberration.) In order to achieve the clarity of vision that has produced so many gorgeous pictures of the planets, stars, and galaxies, a corrective lens was installed on the Hubble in 1993. The Corrective Optics Space Telescope Axial Replacement (COSTAR)[12] is designed to exactly cancel out the spherical aberration of the primary mirror. Cameras mounted on the Hubble in recent years have been built with internal corrective optics that serve the same purpose.

Designing the right corrective lens for the Hubble required a detailed knowledge of the warped shape of the original mirror. Understanding the distortions caused by the misground mirror allowed scientists to effectively subtract them and restore the vision of the telescope. In a similar manner, scientists hope to cancel out the distortions in the cosmic microwave background caused by gravitational lensing in order to reveal the signal of gravity waves.[13]

The lensing effects of the cosmic web of matter can be estimated from a careful examination of the temperature map of the cosmic microwave background—the lensing will smooth out some of the temperature differences in ways that can be statistically extracted—or from the lensing of distant galaxies, as described in the previous chapter. Once these are well understood, the data can be de-lensed—the distorting effects of the cosmic web will be removed as much as possible and the machinations of the Universe in its first few moments of existence will be revealed.

Maybe.

The signal of gravity waves may be too small, remaining stubbornly out of the reach of our detectors. However, there is a good chance that they will be just large enough to impart a noticeable signal in the cosmic microwave background, and the motivation for attempting to dig them out is very high. If gravity waves are found, we can use them to test the theory of inflation, and, in combination with a high-precision temperature map of the cosmic microwave background, we may even be able to point to a particular class of inflation models. The dynamics of the earliest moments of the Universe will be revealed, presenting us with a detailed record of the first fraction of a second.

EPILOGUE

Dark Matter and Dark Energy: Keys to the Next Revolution

The Universe we're exploring today bears little resemblance to the world we thought we inhabited only a short time ago. At the beginning of the twentieth century our concept of the Universe stretched no further than a few hundred thousand light-years into space; protons were not yet part of the scientific lexicon, and time and space were treated as two separate and immutable entities. In the intervening hundred years or so we have expanded our horizons in every possible dimension. We have seen galaxies and quasars billions of light-years from our home planet; traced the history of an evolving and expanding cosmos back to the first moments of space and time; and dissected the atom to reveal a realm of quarks and neutrinos.

What we have learned is amazing. The Universe is 13.7 billion years old, it has an average temperature of just under 3 degrees above absolute zero, and its spatial geometry is flat. The enormous expanse of space that we can see today, filled with hundreds of billions of galaxies, began as an intensely hot, almost infinitely dense soup of energy that has expanded and cooled since the beginning of time and space. Space itself is expanding in a great cosmic stretch that has recently begun to kick it up a notch—the Universe is accelerating. And it is dark. The cosmic inventory is dominated by dark energy (72%) and dark matter (23%); normal matter, which comprises everything we have ever been able to hold in our hands or examine with our instruments, comes in a distant third, contributing only about 5% of everything that is.

The Big Bang theory has been extremely successful in describing the history and evolution of the Universe, and new experiments and observations continue to confirm the basic predictions of this theory. But we do not yet know the answer to the question

What is the Universe made of?

The evidence for dark matter and dark energy has rewritten this question, redirected our efforts to address it—and strengthened the motivation for finding an answer. Dark matter and dark energy cannot be explained by our current model of fundamental physics; deciphering these mysterious substances will require a leap into a new realm of understanding, a theoretical upheaval that will rival the changes brought about by quantum mechanics and general relativity.

General relativity may also lead the way in taking us to this next revolution. Einstein's theory has come full circle, from a radical new concept of space and time, matter and energy, to a practical and powerful tool with which to study these entities. Gravitational lensing—Einstein's Telescope—offers a unique means of exploring the dark sector of the Universe. The ability to detect the presence of mass where no light is found is vital for mapping the distribution of dark matter in galaxies, clusters, and the cosmic web—and for teasing out the effects of dark energy on the expansion and evolution of the Universe.

With the help of gravitational lensing, we have already discovered that dark matter is not composed of MACHOs—dark clumps of normal matter—and determined that the evidence for dark matter cannot be explained by modification of our theories of gravity. Some exotic new particle, as yet undetected, remains the leading contender for the bulk of the matter in the Universe.

We have weighed clusters of galaxies and reconstructed the distribution of matter within these cosmic giants. And we have traced the outline of the dark matter web with weak gravitational lensing—a technique that promises to reveal the influence of dark energy as it begins to dominate the Universe.

We are even beginning to look for ways to probe beyond the beginning of space and time. Gravitational lensing experiments to map the cosmic web of dark matter are an essential first step in our attempts to extract observational hints of inaccessible realms.

THE MULTIVERSE

We have no way of knowing, or testing, what came before the Big Bang, but that has not stopped creative minds from traipsing into this speculative territory. Theoretical cosmologists have argued that if inflation is a valid description of the early Universe, then our observable Universe is only one of an infinite number of universes—part of what has become known as the *multiverse*—and inflation, once started in this mother of all universes, will never end.

The concept of unending, or eternal, inflation was first proposed by Alexander Vilenkin,[1] and implies that our Universe is not all that there is—although we will never be able to see, contact, or explore any of the other universes in the multiverse. Or even have any way to know for sure if they exist.

Consider a universe filled with an energy field that is not at the minimum of its potential, as discussed in Chapter 12. It therefore has a vacuum energy that powers an inflationary expansion of space. This vacuum energy is said to decay when the energy field eventually makes the transition to its minimum level, just as the marble eventually ends up at the bottom of the well in our toy example of inflation (see Illustration 10.5 on page 217). But this decay does not happen everywhere in space at the same time. First one patch of space, then another, will make the transition to the minimum energy state, in a random fashion—the exact mechanism depends on the particular model of inflation. If space were not inflating—expanding at an enormous rate—this story might conclude here. The entire universe would settle down as the energy field rolled to its lowest state at all points in this space, and inflation would end.

Space is inflating, however—so fast that the decay of small regions can't keep pace with the accelerated expansion of space. This is where things get really interesting. When the value of the energy field in one part of this universe reaches its minimum energy, this patch will stop inflating. A baby universe has just been created. From here on out this region is no longer part of the larger parent universe. It takes its relatively small bit of space and starts off on its own evolutionary path, cutting off all further contact with the rest of the original universe. In our Universe, this corresponds to the end of inflation and the beginning of the standard Big Bang expansion. If this theory is correct, we broke off from a larger Universe at this point—and have no way ever to reach it again.

Meanwhile, the parent universe continues to inflate. Until another patch decays and pops out another baby universe. And another . . . and another. As volumes of space are pinched off to form these offspring, the parent keeps expanding. The rate at which space is subtracted from the parent universe to spawn the creation of baby universes is more than offset by the rate at which new volumes of space are created by the expansion. In this scenario, once it is started, inflation will continue forever.

And our Universe may be just one of many.

THE NEXT REVOLUTION

From the multiverse to gravity waves to dark matter and dark energy, the data and theories at the forefront of cosmology today are energizing this incredibly exciting era in the study of the Universe. Our comprehension of the cosmos has reached almost inconceivable new realms, and even more breathtaking is the recognition that there is so much more to be discovered. We don't yet know what the Universe is made of, nor do we truly understand the nature of space and time. More important, we have no idea what new conceptual dimensions will be revealed as we try to solve the mysteries that are currently in front of us.

The search for dark matter and dark energy is certain to overthrow our most basic understanding of the cosmos, and gravitational lensing will continue to play a lead role in this search. The data that is gathered with Einstein's Telescope will be combed for hints of new particles, new theories of gravity, new dimensions of spacetime, and even new theories of the underlying fabric of space and time. We don't yet know the nature of this next revolution or how far from our current picture of the Universe it will take us—but all signs are pointing to something very different from anything we have ever yet imagined.

Notes

The notes, books, and Web sites cited in these notes are accessible to a general audience; the journal papers referred to are usually original scientific works and are marked with an asterisk (*) to indicate a technical reference.

The list of scientific references on gravitational lensing is large and growing, and the sources cited here by no means comprise a complete or comprehensive bibliography on the subject. Rather, I have tried to include references that may serve as a starting point for those who wish to dig deeper into specific areas of the science or the history of this field. In many cases, additional references can be found within these sources.

CHAPTER 1

1. Early in the twentieth century, certain experiments appeared to violate conservation of energy. In the radioactive decay of an unstable nucleus known as beta decay, a neutron in the nucleus is converted into a proton, and an electron comes zinging out (these high-speed electrons were first known as beta radiation). The existence of the neutron was not yet known, but the bottom line was still fairly straightforward: the energy of the emitted electron should have been essentially equal to the mass difference of the nucleus before and after the decay. (The electron should have been carrying away the energy produced in the transformation of the neutron into a proton.) Instead, it never had quite enough—a little bit of energy seemed to disappear. To get out of this conundrum, in 1930 Wolfgang Pauli postulated the existence of a new particle, with no electric charge and little or no mass, which would be produced along with the electron and then zip off unseen, taking with it the missing energy. In 1933, Enrico Fermi dubbed this new particle the neutrino and wrote down a successful theory of beta decay that

included the neutrino. Twenty-three years later, neutrinos were officially added to the particle zoo when two physicists, Fred Reines and Clyde Cowan, detected the neutrinos emitted by a nuclear power plant in Savannah River, South Carolina. More recently, neutrinos were determined to have tiny masses (at least 500,000 times smaller than that of an electron).

2. The Standard Model of particle physics organizes matter particles into "families"—particles within a family tend to interact most often with each other. Each family contains two quarks, an electron-like particle, and its companion neutrino. For reasons that we don't yet understand, there are not one, but three families of basic particles. The discovery of this strange redundancy began in 1937 when a particle was found that seemed to be identical to an electron, except that it was 200 times heavier. This particle, the muon, was the first member of a second family of particles, which also contains the charm quark, the strange quark, and the muon neutrino. The third family of particles—top quark, bottom quark, tau, and tau neutrino—was experimentally completed with the detection at Fermilab of the top quark in 1995 and the tau neutrino in 2000.
3. This quote can be found in S. Mitton, *Conflict in the Cosmos: Fred Hoyle's Life in Science* (Washington DC: Joseph Henry Press, 2005), 128. See also Hoyle's 1950 BBC lectures, *The Nature of the Universe* (New York: Harper & Brothers, 1950), 119.
4. *E. Hubble, "A Relation between Distance and Radial Velocity among Extra-galactic Nebulae," *Proceedings of the National Academy of Sciences, USA* 15, no. 3 (March 15, 1929), 168–173.
5. These observations include data from the WMAP microwave background experiment—*D. N. Spergel et al., "Three-Year Wilkinson Microwave Anisotropy Probe (WMAP) Observations: Implications for Cosmology," *Astrophysical Journal Supplement Series* 170 (June 2007), 377. Data from current measurements of the distance and recession velocity of external galaxies, similar to Hubble's observations, have a precision of about 10%.
6. *For a review of the data, see B. Fields and S. Sarkar, "Big Bang Nucleosynthesis," *Review of Particle Physics 2006*, available at http://pdg.lbl.gov/2007/reviews/bigbang nucrpp.pdf.
7. *A. Penzias and R. Wilson, "A Measurement of Excess Antenna Temperature at 4080 Mc/s," *Astrophysical Journal* 142 (July 1, 1965), 419–421.
8. *F. Zwicky, "Die Rotverschiebung von extragalaktischen Nebeln," *Helvetica Physica Acta* 6 (1933), 110. Note that Zwicky's published results actually stated that there must be 400 times more dark matter than luminous matter; he was using Hubble's value for the expansion rate of the Universe—H_0 = 558 kilometers per second per megaparsec, or (km/s)/Mpc—which is about eight times higher than the current value for this parameter. For a review of the early history of dark matter, see S. van den Bergh, "The Early History of Dark Matter," *Publications of the Astronomical Society of the Pacific* 111 (1999), 657.

9. *S. Smith, "The Mass of the Virgo Cluster," *Astrophysical Journal* 83 (1936), 23.
10. *V. C. Rubin and W. K. Ford, "Rotation of the Andromeda Nebula from a Spectroscopic Survey of Emission Regions," *Astrophysical Journal* 159 (1970), 379.
11. Scientists often use the term *baryonic matter* when talking about the normal-matter contribution to the mass budget of the Universe. Baryons include particles composed of three quarks such as protons and neutrons. Electrons, muons, tau particles, and neutrinos are known as *leptons*. Most of the normal matter in the Universe today is in the form of protons, neutrons, electrons, and neutrinos, and the total mass in these particles is dominated by the mass of the protons and neutrons (the electron and neutrino masses are much smaller). Thus, the mass in baryons and the mass in normal matter are essentially equivalent.
12. From the television documentary *Cosmos* (1980).
13. This result was announced in *P. de Bernardis et al., "A Flat Universe from High-Resolution Maps of the Cosmic Microwave Background Radiation," *Nature* 404, no. 6781 (April 27, 2000), 955. For more information about the Boomerang experiment see http://cmb.phys.cwru.edu/boomerang. Results obtained by MAXIMA, another balloon-borne CMB detector, were announced about two weeks after Boomerang's results.
14. The quest to find a unified theory that incorporates both quantum mechanics and general relativity has occupied a large fraction of the brainpower in physics since Einstein, and it remains a hotly debated issue. Scientists have developed a quantum theory of particles and their interactions under the electromagnetic and nuclear forces, but attempts to quantize the force of gravity in a similar manner (which is essential for understanding how gravity behaves over very tiny distances, for example) have not yet been successful—at least, not without the addition of extra dimensions to spacetime. String theory is part of this effort, but it is still a work in progress.
15. *For the High-z Supernova Search Team, see A. Riess et al., "Observational Evidence from Supernovae for an Accelerating Universe and a Cosmological Constant," *Astronomical Journal* 116 (1998), 1009; for the Supernova Cosmology Project, see S. Perlmutter et al., "Measurements of Omega and Lambda from 42 High-Redshift Supernovae," *Astrophysical Journal* 517 (1999), 565. Brian Schmidt originally formed the High-z Supernova team in 1994 while at the Harvard-Smithsonian Center for Astrophysics, moving to the Australian National University in 1995.

    Note that in order to measure the acceleration of the Universe, it is not necessary to know the absolute value of the expansion rate (the Hubble Constant), only the change in this rate over time. The two supernova teams measured the change in the apparent brightness of supernovae as a function of redshift.
16. *Science Magazine* 282 (December 1998). The term *dark energy* first appeared in a *preprint by D. Huterer and M. Turner, *Physical Review* D60 (1999), 08301.

CHAPTER 2

1. See the discussion in chapter 6 of R. Kolb, *Blind Watchers of the Sky* (Reading, MA: Addison-Wesley, 1996).
2. This instability was first shown by *A. S. Eddington, "On the Instability of Einstein's Spherical World," *Monthly Notices of the Royal Astronomical Society* 90 (1930), 668 (assuming isotropic and homogeneous spatial perturbations).
3. Instead of exactly the same orbit being traced out each time around the Sun, the orbit shifts a bit so that, viewed over time, the path looks more like the pattern produced by a Spirograph toy.
4. A. Pais, *Subtle Is the Lord: The Science and the Life of Albert Einstein* (New York: Oxford University Press, 1982), 253. This excellent book provides a detailed history of Einstein and his work, including a fairly sophisticated discussion of the science.
5. Einstein's 1905 papers can be found in J. Stachel, ed., *The Collected Papers of Albert Einstein*, vol. 2, *The Swiss Years: Writings, 1900–1909* (Princeton, NJ: Princeton University Press, 1989). See also Pais, *Subtle Is the Lord*, for details.
6. Pais, *Subtle Is the Lord*, 250; a slightly different translation appears in J. Stachel, *Collected Papers*, vol. 8, *The Berlin Years: Correspondence, 1914–1918* (Princeton, NJ: Princeton University Press, 1998), 167.
7. We can use the equivalence principle and the example of an accelerating rocket ship to understand gravitational redshifting. Consider an experiment in which a light beam is emitted at the back of the ship and detected at the front. Because the ship is accelerating (its speed is increasing over time), the ship is moving faster when the light is detected than when it was emitted. The front of the ship is effectively moving (away) with respect to the light source, and thus the light will be "Doppler shifted"—lowered in frequency—when it hits the detector, just as the sound of a train whistle is shifted to a lower pitch as the train moves away from you. Because the experiment conducted on board the accelerated rocket ship is equivalent to the same experiment conducted in a gravitational field, the same shift in frequency (a lower frequency corresponds to a decrease in the energy of the light) will be seen in light moving through a gravitational field. Gravitational redshifting was first measured by *R. V. Pound and G. A. Rebka, "Apparent Weight of Photons," *Physical Review Letters* 4 (April 1960), 337–341.
8. Interestingly, this answer had been worked out over 100 years earlier by Johann Georg von Soldner—*"Über die Ablenkung eines Lichtstrahls von seiner geradlinigen Bewegung durch die Attraktion eines Weltkörpers, an welchem er nahe vorbeigeht," *Berliner Astronomisches Jahrbuch* (1804), 161—who used Newtonian gravity to calculate how much a light ray would be bent by the Sun, assuming that light particles had some small but negligible mass. Einstein was unaware of this work when he published his early papers on light bending.

CHAPTER 3

1. G. Gamow, *My World Line* (New York: Viking Press, 1970).
2. *R. Caldwell, M. Kamionkowski, and N. Weinberg, "Phantom Energy: Dark Energy with w<-1 Causes a Cosmic Doomsday," *Physical Review Letters* 91 (2003), 071301.

CHAPTER 4

1. Letter from Einstein to George Hale, October 14, 1913, in J. Stachel, ed., *The Collected Papers of Albert Einstein*, vol. 5, *The Swiss Years: Correspondence, 1902–1914* (Princeton, NJ: Princeton University Press, 1995), 356.
2. K. Hentschel, *The Einstein Tower: An Intertexture of Dynamic Construction, Relativity Theory, and Astronomy* (Stanford, CA: Stanford University Press, 1997); "Our Astronomical Column," *Nature* 94 (1914), 65–66; K. Hentschel, "Erwin Finlay Freundlich and Testing Einstein's Theory of Relativity," *Archive for History of Exact Sciences* 47 (1994), 143.
3. For an overview of the 1919 eclipse, see P. Coles, "Einstein, Eddington and the 1919 Eclipse," in *Historical Development of Modern Cosmology*, edited by V. J. Martínez, V. Trimble, and M. J. Pons-Bordería, Astronomical Society of the Pacific Conference Series 252 (San Francisco: Astronomical Society of the Pacific, 2001), 21. For a report closer to the source, see A. S. Eddington, *Space, Time and Gravitation* (Cambridge: Cambridge University Press, 1921); or *F. W. Dyson, A. S. Eddington, and C. Davidson, "A Determination of the Deflection of Light by the Sun's Gravitational Field, from Observations made at the Total Eclipse of May 29, 1919," *Philosophical Transactions of the Royal Society of London* 220 (1920), 291.
4. See M. Stanley, "An Expedition to Heal the Wounds of War: 1919 Eclipse and Eddington as Quaker Adventurer," *Isis* 94 (2003), 57, for an interesting discussion of Eddington and the claims that he "fudged" the eclipse data.
5. London *Times*, November 7, 1919; see also A. Pais, *Subtle Is the Lord: The Science and the Life of Albert Einstein* (New York: Oxford University Press, 1982), 307.
6. A. S. Eddington, *Report on the Relativity Theory of Gravitation* (London: Fleetway Press, 1918), 54.
7. A. S. Eddington, *Space, Time and Gravitation* (Cambridge: Cambridge University Press, 1921).
8. *O. Chwolson, "Über eine mögliche Form fiktiver Doppelsterne," *Astronomische Nachrichten* 221 (1924), 329.
9. Since Chwolson's 1924 paper (see note 8) is the first published mention of this phenomenon, some have argued that the ring of light should be called a "Chwolson ring."
10. See J. Renn, T. Sauer, and J. Stachel, "The Origin of Gravitational Lensing: A Post-

script to Einstein's 1936 *Science* Paper," *Science* 275 (1997), 184, and references therein. *The 1936 Einstein paper can be found in the journal *Science*, vol. 84, p. 506.

11. *F. Zwicky, "Nebulae as Gravitational Lenses," *Physical Review* 51 (1937), 290; "On the Probability of Detecting Nebulae Which Act as Gravitational Lenses," *Physical Review* 51 (1937), 679.
12. *D. Walsh, R. F. Carswell, and R. J. Weymann, "0957+561 A, B: Twin Quasi-stellar Objects of Gravitational Lens?," *Nature* 279 (1979), 381. See also D. Walsh, "0957+561: The Unpublished Story," in *Gravitational Lenses: Proceedings of a Conference Held at the MIT in Honour of Bernard F. Burke's 60th Birthday*; Lecture Notes in Physics (Berlin: Springer, 1989), 11–22.
13. A few assumptions have been made in determining this angle: (a) the size of the object is small compared to the closest point of approach, and (b) the bending angle is small. These assumptions are reasonable for many of the applications of gravitational lensing that we will discuss.

    To calculate what we would actually observe—for example, the shift in the position of a star behind the Sun during an eclipse, we also need to include the distances between the observer, lens, and light source. The shift—the angular difference $\delta\phi$ between the apparent position of the lensed light source and its position in the absence of the lens—is a function of the mass M of the lens, the impact parameter b (how close the light passes to the lens), the distance between observer and source D_S, and the distance between lens and source D_{LS}.

$$\delta\phi = \frac{4GM}{bc^2}\frac{D_{LS}}{D_S}$$

    For more technical details on gravitational lensing, see for example, *S. Mollerach and E. Roulet, *Gravitational Lensing and Microlensing* (Singapore: World Scientific, 2002), and J. Wambsganss, "Gravitational Lensing in Astronomy," www.livingreviews.org/lrr-1998-12.

14. *J. N. Hewitt et al., "Unusual Radio Source MG1131+0456: A Possible Einstein Ring," *Nature* 333 (1988), 537. The observed radius (R_E) of an Einstein ring depends on the mass of the lens (M) and the distances between the observer, lens, and source:

$$R_E = \sqrt{\frac{4GM}{c^2}\frac{D_L D_{LS}}{D_S}}$$

    D_L is the distance from the observer to the lens; D_S, from observer to source; and D_{LS}, between lens and source. G is the gravitational constant and c is the speed of light.

15. Galaxies (and any lens that does not have a point of infinite density) always produce an odd number of images, although the one that lies near the center of the lens is usually very faint or hidden by the lens itself, so we see two or four images.

CHAPTER 5

1. The disk also contains gas, and the motion of the gas in a galaxy is also used to trace the mass.
2. We know that stars and planets exist and can estimate their numbers from the density of these objects detected so far. These counts indicate that there are not enough planets or dim stars to account for the Galactic dark matter. However, these estimates could be way off. Faint objects are hard to detect, and thus we can see only those that are extremely close to us. This very local sample may not be sufficient to extrapolate to a much larger population. Furthermore, estimates of the numbers of very faint stars assume that the trends that have been observed for brighter stars continue to the faintest stars. The most common mass for stars in our Galaxy is about one-third the mass of the Sun. There are fewer stars with a mass twice that of the Sun; fewer still with a mass 10 times that of the Sun; and so on. Likewise there are fewer stars with one-fourth the mass of the Sun; fewer still with a mass one-sixth that of the Sun. So, as we look for fainter and fainter stars, we find smaller and smaller numbers of them. If we assume that this decrease in number continues, there cannot be enough very dim stars to account for the dark matter.

   There are also limits on the amount of mass that can be in normal matter, which in turn put constraints on the amount of mass that can be hidden in faint stars and planets (which are composed of normal matter). As discussed in Chapter 1, current observations of the cosmic microwave background (CMB) and our understanding of the creation of the light elements preclude a Galaxy filled with normal matter. However, at the time MACHO searches were being designed, the constraints on normal matter were less restrictive.
3. Black holes that are created in supernovae are known as *stellar black holes*. There are also *supermassive black holes* (with masses millions of times the mass of the Sun and thus way too heavy to be MACHOs) and black holes formed in the very early Universe, known as *primordial black holes*. Primordial black holes evade the limits on normal matter from nucleosynthesis; however, their existence is still speculative and the mechanism for their formation is under debate.
4. Otherwise (a) we would see the stars and (b) there would be more mass in these stars than in their planets (just as the mass of the Sun constitutes most of the mass in the Solar System).
5. *B. Paczynski, "Gravitational Microlensing by the Galactic Halo," *Astronomical Journal* 304 (1986), 1.
6. *A. Einstein, "Lens-like Action of a Star by the Deviation of Light in the Gravitational Field," *Science* 84 (1936), 506.
7. The MACHO collaboration had dedicated use of the 1.3-meter telescope at the Mount Stromlo Observatory in Canberra, Australia; the OGLE and EROS teams traveled to

observatories in Chile. EROS initially mounted two searches, both at the La Silla Observatory in Chile. They used a 40-centimeter telescope to take images of the same area of the LMC every few hours or days in order to look for stars that changed brightness on these short timescales, and a 1-meter Schmidt telescope to search for lensing events that lasted for weeks or months. EROS upgraded a few years later to a 1-meter telescope with two new digital cameras. The OGLE collaboration started taking data in 1992, initially targeting only stars in the central core (bulge) of the Galaxy in order to look for dark objects in the disk of the Galaxy. They had limited amounts of time on the 1-meter Swope telescope at Las Campanas Observatory in Chile, and in 1996 they installed the 1.3-meter Warsaw telescope, which is dedicated to their use, to search the SMC and LMC in addition to the Galactic bulge.

8. Recall that the MACHO must be almost directly over a star in order to lens it. The MACHO lens is so small that even though we can't separate the light from the two stars, the lens still passes directly over only one of them. The light from the second star is not affected by the lens.
9. The first results were published back-to-back in the journal *Nature* in 1993: *C. Alcock et al., "Possible Gravitational Microlensing of a Star in the Large Magellanic Cloud," *Nature* 365 (1993), 621; *E. Aubourg et al., "Evidence for Gravitational Microlensing by Dark Objects in the Galactic Halo," *Nature* 365 (1993), 623. For more recent results, see also *C. Alcock et al., "The MACHO Project: Microlensing Results from 5.7 Years of LMC Observations," *Astrophysical Journal* 542 (2000), 281; *P. Tisserand, "Limits on the MACHO Content of the Galactic Halo from the EROS-2 Survey of the Magellanic Clouds," *Astronomy & Astrophysics* 469 (2007), 387.
10. See, for example, *K. Griest, "Galactic Microlensing as a Method of Detecting Massive Compact Halo Objects," *Astrophysical Journal* 366 (1991), 412.
11. For a (technical) discussion of some of these possibilities, as well as a discussion of the next-generation SuperMACHO experiment, see *A. Rest et al., "Testing LMC Microlensing Scenarios: The Discrimination Power of the SuperMACHO Microlensing Survey," *Astrophysical Journal* 634 (2005), 1103, and references therein.

CHAPTER 6

1. There's usually a group of holdouts, leaving 10% or so of the mass in protons, accompanied by an equal number of electrons.
2. *J. Michell, "On the Means of Discovering the Distance, Magnitude, &c. of the Fixed Stars, in Consequence of the Diminution of the Velocity of Their Light, in Case Such a Diminution Should Be Found to Take Place in Any of Them, and Such Other Data Should Be Procured from Observations, As Would Be Farther Necessary for That Purpose," *Philosophical Transactions of the Royal Society of London* 74 (1783), 35.
3. For an excellent book on the science and history of black holes, see K. Thorne, *Black*

Holes and Time Warps: Einstein's Outrageous Legacy (New York: W. W. Norton, 1994). See also correspondence between Einstein and Schwarzschild, December 1915 through February 1916, in J. Stachel, ed., *The Collected Papers of Albert Einstein*, vol. 8, *The Berlin Years: Correspondence, 1914–1918* (Princeton, NJ: Princeton University Press, 1998).

4. Thorne, *Black Holes and Time Warps*, 256; J. A. Wheeler, "Our Universe: The Known and the Unknown," *American Scientist* 56 (1968), 1.
5. Normal stars more massive than the Sun are also much brighter than the Sun and would be seen, so this possibility can be eliminated. The only dark objects that might masquerade as black holes are neutron stars. A neutron star can also pull gas off a stellar companion and heat it to X-ray temperatures, but its mass cannot be more than two or three times the mass of the Sun. Thus, a dark candidate must have a mass securely above this limit in order to qualify as a black hole.
6. Thorne, *Black Holes and Time Warps*, 304. This book also contains a discussion of Cygnus X-1, and the Hawking–Thorne wager.
7. *S. Mao et al., "Optical Gravitational Lensing Experiment OGLE-1999-BUL-32: The Longest Ever Microlensing Event—Evidence for a Stellar Mass Black Hole?" *Monthly Notices of the Royal Astronomical Society* 329 (2002), 349; *D. P. Bennett et al., "Gravitational Microlensing Events Due to Stellar-Mass Black Holes," *Astrophysical Journal* 579 (2002), 639. The plots in Illustration 6.3 are based on figures in Bennett et al., 2002.
8. S. J. Edberg, M. Shao, and C. A. Beichman, eds., *SIM PlanetQuest: A Mission for Astrophysics and Planet-Finding* (Pasadena, CA: Jet Propulsion Laboratory, California Institute of Technology, 2005), available at planetquest.jpl.nasa.gov/documents/White Paper05ver18_final.pdf.
9. *A. Wolszczan and D. A. Frail, "A Planetary System around the Millisecond Pulsar PSR1257 + 12," *Nature* 355 (1992), 145. On another note, it's possible to listen to the music of pulsars—check out the Vela pulsar courtesy of the Jodrell Bank Observatory in England: www.jb.man.ac.uk/~pulsar/Education/Sounds/sounds.html.
10. See NASA's PlanetQuest Web site for the latest in extrasolar planets: http://planetquest.jpl.nasa.gov.
11. *I. A. Bond et al., "OGLE 2003-BLG-235/MOA 2003-BLG-53: A Planetary Microlensing Event," *Astrophysical Journal Letters* 606 (2004), L155; *A. Udalski et al., "A Jovian-Mass Planet in Microlensing Event OGLE-2005-BLG-071," *Astrophysical Journal* 628 (2005), L109; *A. Gould et al., "Microlens OGLE-2005-BLG-169 Implies That Cool Neptune-like Planets Are Common," *Astrophysical Journal* 644 (2006), L37; *J.-P. Beaulieu et al., "Discovery of a Cool Planet of 5.5 Earth Masses through Gravitational Microlensing," *Nature* 439 (2006), 437. The plots in Illustration 6.4 were based on figures found in these references. Three more microlensing planets were reported in 2008.
12. W. Schomaker, "Gravity's Lens Yields a New Planet," *Astronomy Magazine*, May 23, 2005, www.astronomy.com/asy/default.aspx?c=a&id=3151.

CHAPTER 7

1. We can't observe and count up every single bit of matter—normal or dark—in the Universe. It's simply far too big. Furthermore, we don't know the size of the entire Universe. It is likely to be much larger in extent than the distance to the farthest objects we can see, and it may even be infinite. (Scientists often speak of the *observable Universe*, which is the volume of the Universe that we can, in principle, observe—the region of spacetime from which light can have traveled to us over the age of the Universe.) Thus, a more reasonable quantity to talk about is the mass density—the amount of mass in a particular volume of space. It's actually the mass or energy *density* that goes into Einstein's equations.

   The most practical way to proceed is to sample a smaller portion of the entire Universe and determine the densities of normal matter and dark matter in this region. If the sampling region is typical of the Universe at large, we will get the matter census we are seeking. Care must be taken to choose a region that is not too small. We know that on small scales the Universe is lumpy—the density of matter (dark or normal) is not the same everywhere in the Universe. The normal-matter density of the Earth is much higher than the normal-matter density of the space between the Earth and the Moon. The density of dark matter in the center of a galaxy is much higher than the density of dark matter in a region of space with no galaxies. More important, the ratio of dark to normal matter varies from one place to another. In the search for dark matter, cosmologists are therefore looking for something smaller than the entire Universe, but large enough to average out the lumps in the relative amounts of dark and normal matter. Clusters of galaxies turn out to be just what we need.
2. Unlike dark matter particles, which have for the most part only gravitational interactions, particles of normal matter can interact with other normal matter via the electromagnetic force. This means that a gas of normal-matter particles can lose energy by radiating light, and this energy loss allows the particles to slow down. As the particles move more slowly they can clump closer and closer together, and the gas can then collapse to form a star or the disk of a galaxy. This process will tend to move the normal matter around on small scales (such as those of stars and galaxies) in ways that the dark matter cannot follow. This difference between dark matter and normal matter results in a different distribution of the two matter components on small scales within a cluster—for example, a star contains much more normal matter than dark matter, and the central regions of galaxies are much richer in normal matter than are their outer parts—but the *overall* composition of the cluster remains the same as it was when the cluster first began to form.
3. See NASA's Web site: www.nasa.gov/audience/forstudents/postsecondary/features/F_NASA_Great_Observatories_PS.html.
4. It is possible for MACHOs within the halo of a cluster galaxy to produce microlensing

of a background quasar. For details and beautiful images, see www.angles.eu.org/meetings/santander_jw.pdf.

5. See, for example, *R. Lynds and V. Petrosian, "Luminous Arcs in Clusters of Galaxies," *Astrophysical Journal* 336 (1989), 1.
6. Ibid.
7. Both stars and quasars are essentially points of light, which cause "diffraction spikes" in the telescope. The spikes are artifacts induced by the mechanical support structure of the telescope, not real features of the stars or quasars.
8. Redshift (a shift in the spectrum of light toward longer wavelengths) is a measure of an object's velocity away from us. Redshift is usually denoted by the letter z and is defined as the fractional change in the wavelength λ of the light:

$$z = (\lambda_{obs} - \lambda_0)/\lambda_0$$

Redshift is also a cosmic distance indicator. In an expanding Universe, the farther away an object is, the faster it moves away from us, and thus the larger the redshift (the exact relationship between redshift and distance is given by Hubble's law, which is explained in the text box where this note is cited—see page 146). The measured redshift is due to a combination of the intrinsic velocity of the galaxy (how fast it's moving around within the cluster) and the recession velocity due to the cosmic expansion. For galaxies well outside our local area, the redshift is due almost entirely to the expansion of the Universe. Galaxies in a cluster 1 billion light-years away will be zipping around within the cluster at speeds on the order of 1 million miles per hour, while the entire cluster is expanding away from us at a speed of almost 50 million miles per hour. Thus, all the cluster galaxies, regardless of how they are moving within the cluster, will have close to the same redshift. (The small scatter in redshift about the mean value is used to estimate the mass of the cluster via the virial method described earlier in the text.)

To be precise, it is the expansion of space itself that results in the redshifting of distance galaxies (see Chapter 10 for more on this)—galaxies are not moving *through* space away from us, but rather *with* space as it expands. For a more technical discussion, see *S. Carroll, *Spacetime and Geometry* (San Francisco: Addison Wesley, 2004), ch. 3.

The relationship between distance and redshift is dictated by cosmology—the expansion history of the Universe—which in turn depends on the contents of the Universe and the measured expansion rate today. The discussion here assumes the favored standard cosmological model (which includes dark matter and dark energy) in translating redshifts (which are what we measure) into light travel time distances, which are commonly used in public presentations of cosmological distances. The *light travel time distance* is essentially the distance light would have traveled in the time since it was emitted *if the Universe were not expanding*. This is not the distance between us and the source of light today, especially at large redshifts. The object that emitted the light continues to expand away from us so that today it is no longer where it was when the light left it.

Scientists use redshifts exclusively, and this is certainly the most accurate way to present the data; here, however, redshifts will continue to be translated into light travel times because it's more compatible with usual notions of distance for most people. The redshift will sometimes also be included, either in the text or in an endnote—readers can use whichever they are more comfortable with. Ned Wright has a wonderful Web site for calculating distance/redshift, available at www.astro.ucla.edu/~wright/CosmoCalc.html; see also E. L. Wright, "A Cosmology Calculator for the World Wide Web," *Publications of the Astronomical Society of the Pacific* 118 (December 2006), 1711–1715.

9. *W. N. Colley, J. A. Tyson, and E. L. Turner, "Unlensing Multiple Arcs in 0024+1654: Reconstruction of the Source Image," *Astrophysical Journal* 461 (1996), L83.
10. See, for example, *V. Springel, C. S. Frenk, and S. D. M. White, "The Large-Scale Structure of the Universe," *Nature* 440 (2006), 1137–1144; *L. King and V. Corless, "Complex Structures in Galaxy Cluster Fields: Implications for Gravitational Lensing Mass Models," *Monthly Notices of the Royal Astronomical Society Letters* 374 (2007), L37.
11. *K. Sharon et al., "Discovery of Multiply Imaged Galaxies behind the Cluster and Lensed Quasar SDSS J1004+4112," *Astrophysical Journal* 629 (2005), L73; *J. Fohlmeister et al., "A Time Delay for the Cluster-Lensed Quasar SDSS J1004+4112," *Astrophysical Journal* 662 (2007), 62.
12. A core creates a caustic in the lens, and a radial arc is produced if a background galaxy lies just inside the caustic line. A central lens region that is highly nonspherical can also produce radial arcs.
13. The mass quoted here includes only the mass interior to the arcs, within 330,000 light-years of the center of the cluster.
14. *I. I. Shapiro, "Fourth Test of General Relativity," *Physical Review Letters* 13 (1964), 789.
15. Clusters are far from simple lenses—massive galaxies within a cluster can have a strong impact on any light that passes close by, and thus affect the arrival timing of the images.
16. *T. Broadhurst et al., "Strong-Lensing Analysis of A1689 from Deep Advanced Camera Images," *Astrophysical Journal* 621 (2005), 53. Especially see the beautiful images in this paper.
17. *G. Soucail, J. P. Kneib, and G. Golse, "Multiple-Images in the Cluster Lens Abell 2218: Constraining the Geometry of the Universe?," *Astronomy and Astrophysics* 417 (2004), L33; also *Á. Elíasdóttir et al., "Where Is the Matter in the Merging Cluster Abell 2218?" eprint arXiv:0710.5636 (2007).
18. For example, the Hubble Ultra Deep Field was obtained from 11.3 days of Hubble Space Telescope observing time. http://hubblesite.org/newscenter/archive/releases/cosmology/2006/12.
19. For a recent example, see *A. Mahdavi et al., "Evidence for Non-hydrostatic Gas from

the Cluster X-ray to Lensing Mass Ratio," *Monthly Notices of the Royal Astronomical Society* 384 (2008): 1567–1574.

CHAPTER 8

1. After freeze-out, the number density of dark matter particles decreases simply as a result of the dilution effect of the expansion of spacetime (more volume, same number of particles).
2. Radio galaxy 1138-262, reported in "Hubble Captures Galaxy in the Making," Hubble Space Telescope (HST) news release STScI-2006-45, October 2006, http://hubblesite.org/newscenter/archive/releases/2006/45. For wonderful images from the Hubble Space Telescope, along with detailed explanations, visit the HST Web site: http://hubblesite.org/newscenter.
3. The Standard Model discussed in Chapter 1 unifies the electromagnetic and weak forces into one "electroweak" force. This part of the theory is known as *quantum electrodynamics*, or *QED*. The quantum version of the strong nuclear force is *quantum chromodynamics*, or *QCD*. Proposed extensions of the Standard Model unite the strong nuclear force and the electroweak force, known as *grand unified theories* (*GUTs*). These theories often incorporate supersymmetry.
4. Spin is very important in the quantum mechanical description of the subatomic world, but it is an entity that is hard to understand in a classical (nonquantum) sense. Every particle has "spin," which comes in only discrete (quantized) values: 0, ½, 1, 3/2, 2, and so on. Particles with integer spin, such as force particles and the proposed Higgs particle, are known as *bosons*; particles with half-integer spin (all matter particles) are known as *fermions*. Ensembles of bosons behave very differently from those of fermions. Bosons can "condense" at extremely low temperatures to form a new state of matter in which all of the bosons behave as a single entity, sometimes called a superatom. Fermions, on the other hand, can never do this, and in fact, they strongly resist being packed too closely together (the Pauli exclusion principle).
5. For a popular-level book on the Higgs particle, see Leon Lederman, *The God Particle* (Boston: Houghton Mifflin, 1993).
6. String theory and theories with extra dimensions of space are discussed in Brian Greene, *The Elegant Universe* (New York: W. W. Norton, 1999); or Lisa Randall, *Warped Passages: Unraveling the Mysteries of the Universe's Hidden Dimensions* (New York: Ecco, 2005).
7. *Supersymmetry* refers not to a specific model, but to an entire suite of models, each of which has many parameters—such as the mass of the particles in the theory, or the strength of their interactions—whose values are not known or specified by the model. Particle experiments are designed to search a range of possible models and parameters.

8. For more on the Large Hadron Collider (LHC) and its current status, see the LHC home page: http://lhc.web.cern.ch/lhc.
9. For a general introduction, including links to many current dark matter experiments, see the Cryogenic Dark Matter Search (CDMS) public outreach Web site, at http://cdms.berkeley.edu/Education/DMpages/index.shtml. For a technical review, see *L. Baudis, "Direct Detection of Cold Dark Matter," in *Proceedings of the 15th International Conference on Supersymmetry and the Unification of Fundamental Interactions, SUSY 2007, July 26–August 1, 2007, Karlsruhe, Germany*, www.susy07.uni-karlsruhe.de/PROC/ProcSUSY07_FINAL.pdf.
10. An introduction to IceCube can be found at www.icecube.wisc.edu/info. For Fermi, see http://fermi.gsfc.nasa.gov.
11. Milgrom's 1983 papers can be found in the **Astrophysical Journal*, vol. 270, beginning on p. 365.
12. In Newtonian gravity, the gravitational force falls off as the inverse square of the distance: $F \propto 1/r^2$. For two point masses m_1 and m_2, $F = Gm_1m_2/r^2$, where G is the gravitational constant.
13. *J. D. Bekenstein, "Relativistic Gravitation Theory for the Modified Newtonian Dynamics Paradigm," *Physical Review D* 70 (2004): 083509.
14. This is a conservative limit. Some analyses find tighter bounds on the neutrino mass consistent with cosmological data, such as WMAP (Wilkinson Microwave Anisotropy Probe) measurements of the CMB combined with measurements of the large-scale structure in the Universe. See, for example, *S. Hannestad and G. Raffelt, "Neutrino Masses and Cosmic Radiation Density: Combined Analysis," *Journal of Cosmology and Astroparticle Physics* 11 (2006), 016.

CHAPTER 9

1. See, for example, *J. Annis et al., *Dark Energy Studies: Challenges to Computational Cosmology*, white paper submitted to the Dark Energy Task Force (2005).
2. *V. Belokurov et al., "Cats and Dogs, Hair and a Hero: A Quintet of New Milky Way Companions," *Astrophysical Journal* 654 (2007), 897. For recent developments, see the news section of the SEGUE Web page, at http://segue.uchicago.edu. SEGUE (Sloan Extension for Galactic Understanding and Exploration) is part of the second phase of the SDSS (SDSS-II), and one of its key goals is mapping the structure of the Milky Way.
3. One of the assumptions in cosmology is that there are no special directions in the Universe, and so far we have found no evidence to indicate that this is not a good assumption. In the case of weak lensing, the galaxies seen in a small patch of the sky in any particular direction are at various distances—that is, they may appear close to each other from our point of view, but in fact be millions or billions of light-years away from each other (some are close to us, some far away). Two galaxies far apart from one

another will be randomly oriented with respect to one another. (On the other hand, correlations between galaxies that are close neighbors must be considered in the analysis of weak-lensing experiments.)

4. *T. Brainerd, R. Blandford, and I. Smail, "Weak Gravitational Lensing by Galaxies," *Astrophysical Journal* 466 (1996), 623; see also the Sloan Digital Sky Survey at www.sdss.org.
5. Early models of galaxy halos were simple and predicted that cold dark matter is arranged in a spherical halo whose density decreases outward from the center as $1/r^2$, where r is the radius. One of the most popular models for galaxy halos today is called the *NFW profile*, named after its developers—Julio Navarro, Carlos Frenk, and Simon White—who extracted its form from their computer simulation of structure formation. In an NFW halo, the mass density falls off slowly in the inner part (the density is inversely proportional to the radius), drops more quickly (as the inverse square of the radius) over the bulk of the halo, finishing with a steeper falloff (as the inverse cube of the radius, $1/r^3$) in the outermost regions. Weak-lensing studies have so far confirmed the basic picture of galaxy halos, with a falloff in density that is roughly consistent with either the NFW profile or the earlier models. The data are not yet up to the task of distinguishing between the two. (But see *R. Mandelbaum et al., "Density Profiles of Galaxy Groups and Clusters from SDSS Galaxy–Galaxy Weak Lensing," *Monthly Notices of the Royal Astronomical Society* 372 [2006], 758.)
6. *P. Natarajan, G. De Lucia, and V. Springel, "Substructure in Lensing Clusters and Simulations," *Monthly Notices of the Royal Astronomical Society* 376 (2007), 180. For a discussion of radial arcs, see *M. Bartelmann and M. Meneghetti, "Do Arcs Require Flat Halo Cusps?" *Astronomy and Astrophysics* 418 (2004), 413.
7. D. Clowe et al., "A Direct Empirical Proof of the Existence of Dark Matter," *Astrophysical Journal Letters* 648 (2006), L109–L113.
8. These collisions also offer an opportunity to test the nature of dark matter. If the dark matter interacts with itself, for example, it may lag slightly behind the galaxies—although not as much as the gas does. Astronomers are searching for other cosmic collisions in the hope of putting new limits on the strength of such self-interactions. These limits, in turn, will constrain models of dark matter particles.
9. *Clowe et al., "Direct Empirical Proof," L109; *M. Bradac et al., "Strong and Weak Lensing United. III. Measuring the Mass Distribution of the Merging Galaxy Cluster 1ES 0657-558," *Astrophysical Journal* 652 (2006), 937. For more images and animations, see the Chandra Web site: http://chandra.harvard.edu/photo/2006/1e0657.

CHAPTER 10

1. For general accounts of the supernova searches, see S. Perlmutter, "Supernovae, Dark Energy, and the Accelerating Universe," *Physics Today*, April 2003, p. 53; and R. Kirsh-

ner, *The Extravagant Universe* (Princeton, NJ: Princeton University Press, 2002). References to the data can be found in Chapter 1, note 15, of this book.

2. Redshifts $z = 0.16$ to 1. The new supernovae observations were compared with data on nearby supernovae in order to calculate the change in the expansion rate. It's worth noting that the *absolute* intrinsic luminosity of Type Ia supernovae is not needed to determine the acceleration of the Universe—as long as these supernovae are standard candles, the change in the expansion rate can be found by measuring the change in the apparent brightness as a function of redshift.
3. This is commonly known as the *inverse square law*. The light intensity falls off as $1/r^2$, where r is the distance from the light source.
4. The Hubble parameter is the expansion rate today, and it is measured in units of velocity over distance (kilometers per second per megaparsec, where one megaparsec is equal to 3.26 million light-years).
5. *E. Hubble, "A Relation between Distance and Radial Velocity among Extra-galactic Nebulae," *Proceedings of the National Academy of Sciences, USA* 15, no. 3 (March 15, 1929), 168–173. See also V. M. Slipher, "Nebulae," *Proceedings of the American Philosophical Society* 56 (1917), 403.
6. *W. Freedman et al., "Final Results from the Hubble Space Telescope Key Project to Measure the Hubble Constant," *Astrophysical Journal* 553 (2001), 47.
7. As discussed in Chapter 1, light travels at a finite speed, which means that the light from distant objects takes a certain amount of time to arrive at Earth. The farther away the object, the longer it takes light from that object to get here, so we are seeing the object not as it is now, but as it was in the past (when the light left the object).
8. *G. Smoot et al., "Structure in the COBE Differential Microwave Radiometer First-Year Maps," *Astrophysical Journal* 396 (1992), L1. Smoot's Nobel lecture, titled "Cosmic Microwave Background Radiation Anisotropies: Their Discovery and Utilization," is available in *Reviews of Modern Physics*—vol. 79 (2007), p. 1349—and contains an extensive list of CMB references.
9. *P. de Bernardis, "A Flat Universe from High-Resolution Maps of the Cosmic Microwave Background Radiation," *Nature* 404 (2000), 955. Boomerang stands for **B**alloon **O**bservations **O**f **M**illimetric **E**xtragalactic **R**adiation **AN**d **G**eophysics.
10. The symbol Ω is shorthand for the density of matter and energy in terms of the critical density needed for a flat Universe, which is about 10^{-29} gram per cubic centimeter. In a flat universe, the total energy density is equal to the critical density and thus $\Omega_{\text{total}} = 1$. The cosmic microwave background data insist on $\Omega_{\text{total}} = 1$; the cluster data limit $\Omega_{\text{matter}} \approx 0.3$; and the supernova data demand dark energy $\Omega_{\text{dark energy}} > 0$. (More specifically, the observed acceleration constrains $\Omega_{\text{dark energy}} - \Omega_{\text{matter}} \approx 0.45$). The data from any two of these three pieces of evidence, considered jointly, point to a Universe whose main component is dark energy. The combination of all three demand that $\Omega_{\text{dark energy}} \approx 0.7$.

11. A wonderful introduction to quintessence by R. Caldwell and P. Steinhardt, two of the proposers of this concept, can be found at http://physicsworld.com/cws/article/print/402 (November 1, 2000). Their original paper with coauthor R. Dave, is *“Cosmological Imprint of an Energy Component with General Equation of State,” *Physical Review Letters* 80 (1998), 1582.
12. *L. Krauss and M. Turner, “Geometry and Destiny,” *General Relativity and Gravitation* 31 (1999), 1453; *R. Caldwell, M. Kamionkowski, and N. Weinberg, “Phantom Energy: Dark Energy with w<-1 Causes a Cosmic Doomsday,” *Physical Review Letters* 91 (2003), 071301.
13. See http://saturn.jpl.nasa.gov/home/index.cfm.
14. The Apache Point Observatory Lunar Laser-ranging Operation (APOLLO) is the newest lunar laser-ranging experiment and has just announced results with a precision of 1 millimeter (*T. Murphy et al., “The Apache Point Observatory Lunar Laser-ranging Operation: Instrument Description and First Detections,” *Publications of the Astronomical Society of the Pacific* 120 (2008): 20–37). The first observatory to send and receive signals was the Lick Observatory in northern California; the MacDonald Observatory in west Texas has been the main U.S. site for the program for several decades.
15. For a popular-level discussion of modified gravity models, see G. Dvali, “Out of Darkness,” *Scientific American*, February 2004, p. 68.

CHAPTER 11

1. For beautiful images and movies, as well as details of the simulation, visit the Millennium Run Web site, at www.mpa-garching.mpg.de/galform/press.
2. Redshift z =127.
3. The Dark Energy Task Force was headed by Rocky Kolb of the University of Chicago and Fermi National Accelerator Laboratory; the final report is available at www.nsf.gov/mps/ast/aaac/dark_energy_task_force/report/detf_final_report.pdf.
4. Scientists are actually looking for (and finding) a kind of bull's-eye pattern, at least in a statistical sense, initially formed in the fluid of the primordial soup. Dark matter comprises the bulk of the mass, but there is also a component of normal matter in the form of plasma, which behaves differently from the dark matter because it is coupled to light. Tiny overdensities in the soup initially start out with more of everything—dark matter, normal matter (plasma), and light. As discussed in the creation of the cosmic microwave background, in response to the higher pressure in these overdensities the light moves out, dragging the plasma along with it. The light and plasma propagate outward in a spherical shell, moving at the speed of sound. When the light and plasma decouple, the light keeps moving away while the normal matter is left behind in a shell that is overdense in normal matter. Galaxies eventually form in overdense regions—both the cen-

tral region of the original overdensity, and also the shell. For a slightly technical, but very descriptive and visual, discussion of this phenomenon, see D. Eisenstein's Web page, at http://cmb.as.arizona.edu/~eisenste/acousticpeak/acoustic_physics.html.

5. *D. Bacon, A. Refregier, and R. Ellis, "Detection of Weak Gravitational Lensing by Large-Scale Structure," *Monthly Notices of the Royal Astronomical Society* 318 (2000), 625; D. Wittman et al., "Detection of Weak Gravitational Lensing Distortions of Distant Galaxies by Cosmic Dark Matter at Large Scales," *Nature* 405 (2000), 143; N. Kaiser, G. Wilson, and G. Luppino, "Large-Scale Cosmic Shear Measurements," eprint arXiv:astro-ph/0003338 (2000); L. Van Waerbeke et al., "Detection of Correlated Galaxy Ellipticities from CFHT Data: First Evidence for Gravitational Lensing by Large-Scale Structures," *Astronomy and Astrophysics* 358 (2000), 30.
6. A second method, which was developed only recently, leapfrogs over the slicing technique. It inputs all of the galaxies and all of their distances into an enormous database and lets a computer sort out the best three-dimensional model that is consistent with all of this data. The underlying concept is the same—the lensing of a distant galaxy is most influenced by the section of the dark matter web roughly halfway between the galaxy and our telescopes—it's just a bit hard harder to visualize.
7. See www.nlm.nih.gov/research/visible/visible_human.html.
8. R. Massey et al., "Dark Matter Maps Reveal Cosmic Scaffolding," *Nature* 445 (2007), 286.
9. See the Dark Energy Task Force final report, at www.nsf.gov/mps/ast/aaac/dark_energy_task_force/report/detf_final_report.pdf.
10. See www.esa.int/science/planck.
11. See http://pole.uchicago.edu.
12. See, for example, B. Bassett, R. Nichol, and D. Eisenstein, "Sounding the Dark Cosmos," *Astronomy & Geophysics* 46 (2005), 5.26.
13. For summaries of these projects, see the Dark Energy Task Force final report, at www.nsf.gov/mps/ast/aaac/dark_energy_task_force/report/detf_final_report.pdf.
14. Large Synoptic Survey Telescope: www.lsst.org.
15. Joint Dark Energy Mission: http://universe.nasa.gov/program/probes/jdem.html.
16. Square Kilometer Array: www.skatelescope.org.
17. The intrinsic shapes of these distant radio sources are less well-known than those of their optical counterparts, but the additional resolving power of a radio survey more than compensates for this handicap.

CHAPTER 12

1. See the discussion in Rocky Kolb, *Blind Watchers of the Sky* (Reading, MA: Addison-Wesley, 1996). See also *G. Gamow, "The Physics of the Expanding Universe," *Vistas in Astronomy* 2 (1956), 1726.

2. Tiny variations in the temperature, at the level of about one part in 100,000, do exist—these are the predicted variations that correspond to matter overdensities in the early Universe and seed the formation of galaxies and clusters. The puzzle is not the existence of the microwave background, nor the minuscule temperature variations, but the fact that the temperature is essentially the same everywhere throughout the observable Universe.
3. In order to determine the size of the Universe, it's essential to include the expansion and be careful with definitions. If there were no expansion, then since the Universe is 13.7 billions years old, the most distant sources we would be able to see would be 13.7 billion light-years away. But the Universe has been expanding and thus the distance *today* to the farthest objects we can see—those that emitted light 13.7 billion years ago—is much larger (the space between us and the object has been growing during this time), such that it is now about 46 billion light-years away.
4. A. Guth, *The Inflationary Universe* (Reading, MA: Addison-Wesley, 1997).
5. In 1999, Alan Guth generously lent his notebook to the Adler Planetarium & Astronomy Museum in Chicago, where is it on display as part of the cosmology exhibit.
6. Earlier inflation corresponds to a higher energy of the Universe.
7. See the Nobel Prize press release at http://nobelprize.org/nobel_prizes/physics/laureates/1993/press.html.
8. See www.ligo-la.caltech.edu.
9. See http://lisa.nasa.gov. Note that current NASA budget cuts may delay the launch of LISA (see the report by the Committee on NASA's Einstein Program titled *NASA's Beyond Einstein Program: An Architecture for Implementation* [Washington DC: National Academies Press, 2007], www.nap.edu/catalog.php?record_id=12006#toc).
10. See Y. D. Takahashi, "Cosmic Microwave Background Polarization: The Next Key toward the Origin of the Universe," at http://bolo.berkeley.edu/~yuki/CMBpol/CMBpol.htm, for graphics and animations of CMB polarization (somewhat technical).
11. The first detection of CMB polarization was presented by the DASI collaboration at the COSMO-02 conference on September 19, 2002; the results were published in *J. Kovac et al., "Detection of Polarization in the Cosmic Microwave Background Using DASI," *Nature* 420 (2002), 772.
12. See http://hubble.nasa.gov/missions/sm1.php.
13. Gravitational lensing will also affect the observations of other distant sources, including high redshift supernovae.

EPILOGUE

1. See A. Guth, *The Inflationary Universe* (Reading, MA: Addison-Wesley, 1997), ch. 15, for an excellent discussion of this theory. See also *A. Vilenkin, "Birth of Inflationary Universes," *Physical Review D* 27 (1983), 2848.

Illustration Acknowledgments

Original artwork and illustrations (except where otherwise noted) by José Francisco Salgado. For these acknowledgments, note the following abbreviations: ESA = European Space Agency; NASA = National Aeronautics and Space Administration; STScI = Space Telescope Science Institute.

CHAPTER 1

1.3 Millennium Run simulation; V. Springel and the Virgo Consortium, Max Planck Institute for Astrophysics, Garching, Germany; NASA, ESA, K. Sharon (Tel Aviv University), and E. Ofek (Caltech); the Hubble Heritage Team AURA/STScI/NASA.
1.4 The Boomerang Collaboration.

CHAPTER 2

2.1 From the collection of the Adler Planetarium & Astronomy Museum, Chicago, Illinois.
2.3 ESA – Pedro Duque.

CHAPTER 3

3.2 José Francisco Salgado and Julieta Aguilera.

CHAPTER 4

4.2 Reproduced with permission from the Einstein Archives at the Hebrew University of Jerusalem.

4.3 George Rhee, NASA/STScI, and William Keel.

4.5 J. N. Hewitt and E. L. Turner. This image was generated with data from telescopes of the National Radio Astronomy Observatory, a National Science Foundation facility managed by Associated Universities, Inc.

4.7 Lensing program courtesy of Pete Kernan.

4.8 NASA and ESA.

CHAPTER 5

5.1 COBE/DIRBE – NASA/Goddard Space Flight Center.

5.2 NASA and ESA.

CHAPTER 6

6.6 Brad Miller.

CHAPTER 7

7.1 W. N. Colley and E. Turner (Princeton University), J. A. Tyson (Bell Labs, Lucent Technologies), and NASA.

7.2 ESA, NASA, K. Sharon (Tel Aviv University), and E. Ofek (Caltech).

7.3 Space Telescope Science Institute, Ann Feild.

CHAPTER 10

10.1 NASA, ESA, the Hubble Key Project team, and the High-z Supernova Search Team.

10.4 With acknowledgment to Perlmutter et al. and the Supernova Cosmology Project, who presented an early version of this plot.

CHAPTER 11

11.1 Millennium Run simulation; Volker Springel and the Virgo Consortium, Max Planck Institute for Astrophysics, Garching, Germany.

11.2 S. Colombi and Y. Mellier (Institut d'Astrophysique de Paris, CNRS).

11.3 B. Jain, U. Seljak, and S. White (2000).

11.4 NASA, ESA, and R. Massey (Caltech).

CHAPTER 12

12.3 K. Thorne (Caltech) and T. Carnahan (NASA/Goddard Space Flight Center).

COLOR ILLUSTRATIONS

1 NASA, ESA, A. Bolton (Harvard-Smithsonian Center for Astrophysics, CfA), and the Sloan Lens ACS Survey (SLACS) team. Reprinted by permission from Macmillan Publishers, Ltd: *Nature* 365 (1993), 621.

2 NASA, ESA, J. Blakeslee, and H. Ford (Johns Hopkins University).

3 C. Alcock and the MACHO collaboration.

4 W. N. Colley and E. Turner (Princeton University), J. A. Tyson (Bell Labs, Lucent Technologies), and NASA.

5 ESA, NASA, K. Sharon (Tel Aviv University), and E. Ofek (Caltech).

6 NASA, N. Benitez (Johns Hopkins University [JHU]), T. Broadhurst (Racah Institute of Physics/Hebrew University), H. Ford (JHU), M. Clampin (STScI), G. Hartig (STScI), G. Illingworth (University of California Observatories/Lick Observatory), the ACS Science Team, and ESA.

7 NASA, Andrew Fruchter, and the ERO Team (Sylvia Baggett, STScI; Richard Hook, Space Telescope-European Coordinating Facility [ST-ECF]; Zoltan Levay, STScI).

8 *X-ray:* NASA/CXC (Chandra X-ray Observatory)/CfA (Center for Astrophysics)/M. Markevitch et al. *Optical:* NASA/STScI; Magellan/University of Arizona/D. Clowe et al. *Lensing map:* NASA/STScI; ESO WFI (European Southern Observatory Wide Field Imager); Magellan/University of Arizona/D. Clowe et al.

9 Millennium Run simulation; Volker Springel and the Virgo Consortium, Max Planck Institute for Astrophysics, Garching, Germany.

10 NASA/WMAP (Wilkinson Microwave Anisotropy Probe) science team.

Index

Page numbers in *italics* refer to illustrations; page numbers after 271 refer to endnotes.

ABOUT THE AUTHOR

Evalyn Gates is the Assistant Director of the Kavli Institute for Cosmological Physics and a Senior Research Associate at the University of Chicago. The former Director of Astronomy and Vice President for Science & Education at the Adler Planetarium & Astronomy Museum, she received her undergraduate education at the College of William & Mary and earned a PhD in theoretical physics at Case Western Reserve University. Following postdoctoral fellowships at Yale University and the University of Chicago, she spent time as a guest scientist at Fermi National Accelerator Laboratory. Her research focuses on particle cosmology and astrophysics, including dark matter, MACHOs, and white dwarfs.